# Biodynamic, Organic and Natural Winemaking

T0178837

# Biodynamic, Organic and Natural Winemaking

## Sustainable Viticulture and Viniculture

BRITT & PER KARLSSON

Floris
Books

Text by Britt Karlsson
Photography by Per Karlsson
English translation by Roger Tanner, Ordväxlingen AB

First published in Swedish as *Vinet Och Miljön: Ekologiskt,
biodynamiskt och naturligt* by Carlsson Bokförlag, Stockholm in 2012
Translated from the first edition and first published in English
by Floris Books, Edinburgh in 2014
Fourth printing 2021

British Library CIP Data available
ISBN 978-178250-113-8
Printed and bound in Great Britain by Bell & Bain, Ltd

Floris Books supports sustainable forest management
by printing this book on materials made from wood that
comes from responsible sources and reclaimed material

# CONTENTS

Foreword                                             13

1. Organic Wine-Growing – an Overview                19
   Is the wine any different?                         20
   Locally grown and small-scale                     22
   Mind your language!                               26
   Respect for the environment                       26

2. Farming Today                                     29
   The four main groups                              29
   Conventional farming                              29
   Organic farming                                   30
   Sustainable farming                               30
   Biodynamic farming                                31
   Common objectives                                 32

3. History in Brief                                  35
   Crisis in the vineyard                            36
   The post-war years                                36
   Awareness                                         40
   Organic wine-growing                              40

4. How Widespread is Organic Wine-Growing?           47
   Spain and Italy top the list                      47
   France                                            48

Austria and Germany                                          49
The New World                                                49

## 5. Who Goes Organic?                                      55
And why?                                                     56
The environment matters most                                 56
Are organic wines more expensive?                            58

## 6. Organic Wine-Growing                                   61
How does organic differ from conventional?                   64
Good soil                                                    65
Roots matter                                                 67
What does the vine need?                                     68
Weeds and cover crops                                        72
The comeback of the hedge                                    74

## 7. Pests and Diseases                                     79
Fungal diseases                                              79
Other pests                                                  81

## 8. Pest Control by Natural Means                          89
PNPP (*Préparations naturelles peu préoccupantes*)           89
The nettle war                                               89
Approval of organic plant protection products                90
Copper and its alternatives                                  91
Spraying                                                     95

| | | |
|---|---|---|
| | Crosses and hybrids | 98 |
| | Genetically modified grape vines | 99 |
| **9.** | **I Want To Go Organic – How is it Done?** | **101** |
| | Official labelling | 101 |
| | The whole EU procedure | 103 |
| | The USA | 105 |
| | Switzerland | 106 |
| | Non-certification | 107 |
| **10.** | **Biodynamic Wine Production** | **111** |
| | The practical side | 113 |
| | *The preparations* | *113* |
| | *500 – Cow dung in cow horn* | *114* |
| | *501 – Quartz in cow horn* | *115* |
| | *508 – Horsetail* | *116* |
| | *The dynamiser and dynamisation* | *116* |
| | *The compost* | *119* |
| | *Compost turns biodynamic* | *119* |
| | *Preparations 502–507* | *120* |
| | Cosmic rhythm, the moon, and four kinds of day | 122 |
| | The consultants | 124 |
| | Control and rules | 126 |
| | Wine-growers on the practical results of biodynamics | 128 |

## 11. Private Labelling and Control    133

Private organic labels and organisations (a selection)    134

*International*    135

*France*    135

*Austria*    136

*Switzerland*    136

*Italy*    137

*Germany*    137

*California*    137

*Australia*    139

*South Africa*    139

*Chile*    139

*Argentina*    139

*New Zealand*    140

Biodynamic certification labels    140

## 12. The Work Inside the Cellar    145

Why have rules in the cellar?    145

Why add sulphur anyway?    146

Reduce by how much?    149

The disadvantage of reducing sulphur content    151

What is typical?    151

Organic and authentic?    152

## 13. Additives      155

Are additives and interference needed?    157
Allergenic additives    157
Which additives are used, and why?    158
Additives and processes during vinification    159
Other matters    179

## 14. Sustainable Wine-Growing      185

The definition of sustainable    185
How do sustainable wine-growers do things?    187
Sustainability by the Cousinié method    190
Labelling    192
   *France*    *192*
   *Austria*    *194*
   *California*    *194*
   *Chile*    *197*
   *South Africa*    *197*
   *New Zealand*    *198*
   *Australia*    *200*
   *Fairtrade*    *200*
   *Code of conduct for Nordic monopolies*    *201*

15. *Vins Nature* – Natural Wines    **203**
     French natural wines    203
     Italian natural wines    205
     The taste    206

16. The Environmental Big Picture    **209**
     Carbon dioxide emissions    209
     Wine transport    212
     Carbon-neutral wines    213
     Lighter bottles    215
     Natural cork    217
     Solar collector panels    218
     Irrigation    219
     Trapping carbon dioxide    220

Appendix 1: Recommended 'natural wine' producers    225

Appendix 2: A selection of favourite organic and biodynamic wine producers    227

Index    259

# LA CAVE DES VINS BIO

Fontaine
à vin

Coteaux varois
rouge (perles)
Corbières AOC
Bordeaux

Brouilly

## Les Rosés AB

l'expression d'un terroir — millésime

VDP Provence sec          5 €
AOC frais et fruité       7,10 €
VDP Varois typique        5,30 €
AOC Bourgogne rosé
ample et généreux         8,60 €
VDP rosé vieux du Rhône   3,60 €

## Les Festives "Petite Soif"

- rosé (chaint Bordelais)  3,60 €
- blanc (généreux et structuré)

Blancs Monocépage
Marsanne, Viognier, Chardonnay...

# Foreword

Today there are many wine producers for whom awareness of how they treat the environment in and around their vineyards comes naturally. We see this on our wine tours in France, Italy, South Africa, Portugal and other countries. The subject is one which the wine-growers themselves readily bring up. It is also one which our fellow-travellers are often very interested in, and one which can easily trigger discussions and debates.

French studies recently showed large wine-growing districts like Bordeaux and Languedoc-Roussillon to have lower levels of microbiological activity than soil in which other crops are grown. The wine industry has been, and remains, a big user of chemical pesticides. As yet, only a tiny fraction of producers dispense entirely with synthetic chemical products in their vineyards – in other words, work their land organically – but the number is growing day by day.

It was this rapid increase that moved us to take a closer look at the nuts and bolts of going organic, but also at the producers' reasoning, their priorities and the different opinions prevailing about organic growing. And opinions there are in plenty. Not everyone agrees on the best way of safeguarding the environment.

This book is not meant as an argument for organic (or biodynamic or natural) wine production. Instead we explain what all these concepts mean. To have an opinion on whether something is good or bad, you need first of all to understand what it means, and it is those very facts we present in this book. Only too often in debates on the environment, people bandy around various terms and concepts without really understanding their meaning.

Here, as in so many other cases, there are no absolute truths.

Take the following example. The EU has defined organic growing as meaning the non-use of artificial fertilisers and synthetic chemicals in the vineyard. On the other hand, spraying or dusting with copper is permissible. Yet many consider copper to be a worse option than synthetic products. But if you use a synthetic alternative, you cannot call yourself an organic grower. It all boils down to striking balances, drawing lines and, of course – politics. All districts, including the cool, rainy, northern ones, must have a chance of producing organic wine, and so the rules have to allow quite generous latitude. Up to 6 kg copper annually may be used in the organic vineyard. Why 6 kilos? Because in a rainy district it is hard to combat certain fungal diseases with less. And so the line is drawn there, despite large amounts of copper not being good for the soil.

Starting from the 2012 vintage, not only are the grapes organically grown, we also officially have an 'organic wine' label. There are a

host of different techniques and additives which can be used during vinification and storage to give the wine the taste profile the producer wants. The EU has defined what is prohibited if you want to call your wine 'organic'. But this again sometimes feels somewhat arbitrary. Why is flash pasteurisation forbidden while reverse osmosis is not? Why are you allowed to chaptalise (add sugar to) the must in order to raise the alcohol content but forbidden to use vacuum distillation to lower it?

The rules inside the cellar have to suit all kinds of organic wine-growers – the small ones who produce distinctive, personal wines, and those mass-producing cheap, streamlined wines. Organic wines come in all price brackets and guises, just as in the conventional wine trade.

Many organic wine-growers feel that the rules are far too lax and, accordingly, work far more restrictively. So at the end of the day it is still the individual growers who decide what sort of wine they want to make.

As in so many other instances where wine is concerned, one is well advised to look at more than just the label. It is even more important to know the wine's producer and the values they subscribe to.

Outside the EU, the rules may be different, for instance, in the USA. And there are some countries which do not yet have any rules at all.

Organic growing is a new phenomenon which has attracted widespread attention. Things are moving fast, and it has been impossible to give the full picture. We have tried to give as fair a picture as possible of organic, biodynamic, natural and sustainable winegrowing. Everyone believes passionately in their own model, they have many points in common and they are inspired by each other.

We can't really say what is best. It is down to the wine-growers themselves to convince us.

One thing is for sure. There are going to be some very interesting developments over the next few years. Perhaps a good alternative to copper will be found. Perhaps more people will realise they ought to plant grape varieties with greater resistance to various diseases. Perhaps energy and water consumption will prove after all to be the biggest environmental issue where the wine industry is concerned. The next few years will bring an upsurge of debate on every aspect of organic, sustainable, biodynamic and natural methods.

Britt Karlsson
Per Karlsson

# CHAPTER 1

# Organic Wine-Growing – an Overview

Interest in organic and otherwise environmentally friendly wines is running high all over the world today. It has quite newly arisen and is growing steadily among both consumers and importers.

Being organic has, of course, to do with taking good care of the environment, but for many producers organic growing is also about being able to make a better, more interesting wine, a wine of greater originality. The belief is that a wine coming from unsprayed soil can absorb the character which the soil is capable of imparting to it.

Talking to growers out in the wine districts, we notice how more and more of them are giving thought to the environmental impact of their work in the vineyard. One producer after another is 'converting' to organic growing. There are rules of procedure for this kind of growing, both official ones and rules issued by private organisations. Aside from organic, people are also speaking more and more of other, closely related concepts – biodynamic, natural wines, sustainable growing. These are all different things, and we will be addressing all of them.

The rules and regulations are somewhat of a jungle. What is regulated and who actually decides? Do all countries have the same rules, and if not, how do the rules differ from one country to another? Not least, many consumers, are asking themselves if the wines actually turn out better. To that one can only reply: perhaps. After all, quality depends mainly on the producer. Getting fine grapes is not enough. You must also be able to turn them into a good wine in the cellar, and some organic producers, of course, have better or worse wine-makers than others. But is it necessary for the wine to be better? Couldn't a wine score plus points purely by virtue of its environmentally friendly production?

*Nesting box in Le Fraghe vineyard, Bardolino.*

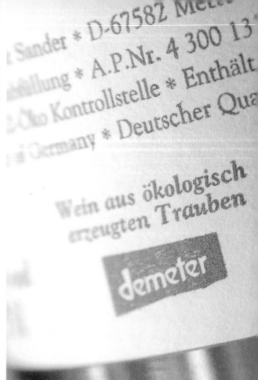

*Cleaning the spraying machine.*

*'Wine made from organically grown grapes.'*

## Is the wine any different?

Does organic growing change the wine at all? The growers themselves claim it does. The acidity improves, they say, the grapes ripen earlier, the balance of the wine improves. It can be hard for the consumer – and the wine expert – to tell what is due, respectively, to organic growing and to the skill of the wine-grower. Some claim to be able to tell if a wine has been made from organically grown grapes, but this is by no means self-evident and is perhaps mostly imagination or autosuggestion.

No other agricultural product is so intimately tied to its *terroir* – its origin – as wine.

The very notion of original character is practically unique to wine. It hardly ever occurs, or at least much less frequently, where vegetables, fruit and cereals are concerned. Now, if the character of wine is affected by the soil the vine grew in, are organic wines alone in having the genuine flavour of their origin? Of course not. It may be tempting to believe that organic wines have a more 'genuine' taste, but no one can yet prove it. And how do you go about proving the character of a geographical origin?

*A Chilean wine certified by IMO Control.*    *Enzymes.*

The designation 'organic wine' is a brand new one in the EU. As from the 2012 vintage, organic growers may call their wines organic instead of just 'wine made from organically grown grapes'. What happened was that in February 2012, the EU reached agreement on the rules governing the production of organic wine in the cellar.

Before that date, the same rules concerning additives and production technique had applied as to conventional wine-growers, that is, there were not any ecological rules as such in the wine cellar. Many organic growers therefore belong to private associations with controls and rules which are stricter than those of the EU. The USA is one of the few countries where it has been possible for several years now to speak of 'organic wines'. On the other hand, very few American wine producers are equal to the requirements this entails.

So are organic grapes grown without any spraying? Only in very few instances. Organic grapes are also sprayed, but with products which in most case are less dangerous to the environment and humans. But there are products, copper not least, which the organic growers are heavily criticised for using.

*Insect counter: biodiversity and numbers are important.*

*Every control organisation is identified by country and number.*

## Locally grown and small-scale

Interest in organic growing is bound up with the reaction we have seen in the past fifty years against large-scale industrialised farming. Consumers are looking for locally produced food products, which may or may not have been grown organically. Many of these are niche products – a kind of luxury article – and not readily available to everybody. Many consumers associate organic growing with smallness of scale and believe (or feel) that it is incompatible with large-scale operations. But there are many examples of large-scale organic growing, as regards both wine and foodstuffs. Organic wine can both be handcrafted on the small family farm and take the form of cheap supermarket wines produced by the big wine companies (Tesco in the UK and Systembolaget in Sweden are big buyers of cheap organic wines).

Not everyone can get hold of locally grown wine. Wines sold in northern Europe have to be carried long distances, either by freighter from the New World or by road or rail from the wine-producing

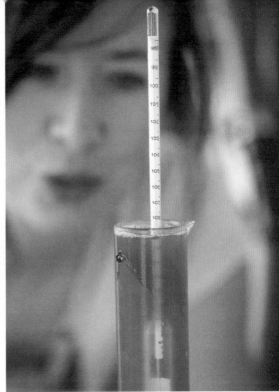

*Organic and biodynamic are different things.*    *A smattering of chemistry comes in useful in the wine cellar.*

countries of Europe. The environmental impact of transport and other handling of wine has now made sustainability a talking point. Which packaging is kindest to the environment – BiB (bag-in-box) or bottle? Which closing device can be recycled? How much do the bottles weigh? Sometimes vineyard employees' working conditions are also factored into the environmental equation. A wine producer has a lot to think about nowadays – weather conditions are no longer the only thing they need worry about.

Demand for organic wines is rising. They attract a lot of press coverage, which is doubtless one reason for the growth of interest. Demand for all organic foodstuffs has skyrocketed in recent years, but the number of people who say they want to shop organically still exceeds the number actually doing so. This may be because the range of organic products is too narrow, prices are too high, people have to track down specialised stores which are too far away, and so on. The spirit may be willing but beset with practical difficulties. Or else people just cannot be bothered to go to that little extra bit of trouble.

*Blue flowers in the Domaine Gayda vineyard, Languedoc.*

## Mind your language!

The environmentally friendly wine scene abounds with jargon: biological, organic, biodynamic, sustainable, natural, integrated and so-called *lutte raisonnée*, to mention but a few. It is not always easy to know what's what and the terms are sometimes carelessly bandied about by producers, consumers and reviewers alike. There are various control agencies monitoring compliance with a diversity of rules. The rules themselves vary from country to country and from one control agency to another. The whole of the 'environmentalist movement' is a bit of a jungle, which makes things difficult for a consumer wanting to know exactly how the grapes have been grown and the wine made.

## Respect for the environment

Basically, it is quite straightforward. Or rather, it ought to be. It is all about respect for the environment. Using products which are not harmful to humans, flora and fauna. But opinions vary as to which methods are best for the environment – the organic or the non-organic. The debate on how our planet will be able to feed an additional three billion people in 2050 is very much concerned with how an organic farm could have the same yield as a conventional one. Wine production is less dependent on high yields – if anything, the opposite applies: a wine-grower aiming for quality limits his yields. Wine is indeed an agricultural product, but not exactly the staff of life. We can manage without it. And so one may ask whether wine-growers will perhaps come under increasing pressure to go organic.

People have planted vines where they want them, and this is not always in the most suitable places. Rain and damp often make synthetic chemical products seem unavoidable in the eyes of wine producers. Putting it drastically, one may ask whether there is any real justification for spraying with products which may be environmentally hazardous, merely in order to produce wine in a region which is really quite unsuitable for it. There are wine-growers who answer in the negative and do not hesitate to use hybrids (crosses between European and American vines) which are resistant, for example, to

*Vine at Château Lafleur, Pomerol.*

downy mildew, a serious fungal disease. Such growers are above all to be found in Belgium and Switzerland. They are not yet numerous, but the example is interesting. Time will tell whether hybrids or perhaps genetic modification (at present unthinkable for any organic grower) can become an option on a larger scale.

France, one of the world's biggest wine producers, has set itself the target of 20 per cent of its agricultural acreage being organically farmed in 2020. The figure at present is below 5 per cent for farming as a whole but 8 per cent for viticulture. Wine-growing is the sector where things are moving fastest. On the other hand, it has some catching up to do, as being the agricultural sector with the highest pesticide use in percentage terms.

Organic food and wine have moved from being niche products which many people did not bother about – shunned, if anything – to a commercial success story. Not many wine-growers bothered about certification a few years ago, because organic growing was no sales argument. That is now changing quickly.

How far can it go? Some believe that before long 50 per cent of French vineyard acreage will be organic. But, as we said, vines sometimes grow in areas with adverse conditions. Too much rain, too much damp. Fungal diseases are the wine-grower's biggest headache, and they are hard to overcome. More research is needed into environmentally friendly products. Nature is full of plants with a potential for helping. But research needs funding, which is less readily forthcoming for natural products than for new synthetic pesticides. But the big agrochemical corporations are putting their house in order and are now also playing a part in research to do with environmentally friendly, sustainable products.

# CHAPTER 2

# Farming Today

## The four main groups

Farming today can be based on a variety of principles, by far the commonest being 'conventional' farming, which is paramount worldwide. Then there are three main groups which, in their various ways, claim to be best at respecting the environment: organic, sustainable and biodynamic farming.

## Conventional farming

This is 'ordinary' farming of the kind which we all grew up with and which has existed for over fifty years. Conventional farming has developed through the use of artificial fertiliser and synthetic chemical pesticides, aimed at achieving big and reliable harvests. Today this farming is also changing and is trying to limit such harmful effects as nitrogen leakage and eutrophication of lakes by using some of the tricks of the environmentalist lobby, added to which, the EU has banned a large number of more or less dangerous products. Something like 700 active substances were proscribed during the big clean-up between 1993 and 2009.

*Vat of organic champagne at Champagne Marguet.*

## Organic farming

*Biologique* or just *bio* in French, *biologico* in Italian, *ecológico* in Spanish.

Growing without the use of synthetic chemical products or artificial fertiliser. There are rules laying down what may be used in the way of nourishment and soil improvement. Genetically modified products are prohibited.

The EU's general definition of organic farming runs as follows:

Organic production is an overall system of farm management and food production that combines best environmental practices, a high level of biodiversity, the preservation of natural resources, the application of high animal welfare standards and a production method in line with the preference of certain consumers for products produced using natural substances and processes. The organic production method thus plays a dual societal role, where it on the one hand provides for a specific market responding to a consumer demand for organic products, and on the other hand delivers public goods contributing to the protection of the environment and animal welfare, as well as to rural development.

Organic certification requires a three-year conversion period and is followed by annual inspections.

## Sustainable farming

In French: *agriculture durable, culture raisonnée, lutte raisonnée, lutte intégrée.*

The philosophy of sustainable farming is to respect the environment by using synthetic chemical products only when absolutely necessary and then in small doses. Sustainable farming also takes into account the safety and wellbeing of employees, for instance through training in the avoidance of injuries to do with the use of equipment and chemical products. The environment is addressed in broad terms through waste management, energy saving, transport operations and so on.

Many people regard sustainable farming as a golden mean. Sustainable growers pursue in many ways the same objectives as organic growers, even though sometimes they completely disagree. The sustainable growers simply do not believe in the possibility of dispensing with chemical spraying entirely, nor perhaps do they believe it desirable, given that the ecological alternatives are not always environmentally harmless.

For some, sustainable wine-growing is the only true method, while for others it is a staging post on the road to becoming organic. There are organisations with rules (or rather, recommendations), but few wine producers as yet are registered. However, the group is expanding steadily. It is hard to say how many are working this way or exactly how they are working. Many profess and call themselves sustainable – no one nowadays wants to be anything but. Growers want to be friends of the environment, and the majority see no point in spraying the vineyard unnecessarily. On the other hand, they have no desire to risk losing parts of the harvest on account of bad weather. It is often hard to know what they actually do.


## Biodynamic farming

French: *biodynamique*, Italian: *biodinamica,* and Spanish: *biodinámica.*

Briefly, biodynamic farming is organic farming plus the anthroposophist Rudolf Steiner's principles on how to restore strength and vitality to the soil. The various 'preparations' made from various plants and either added to the compost or sprayed on the vineyard and vines are of cardinal importance. The biodynamic 'Sowing And Planting Calendar' must be followed as closely as possible.

Biodynamic wine-growers are regulated by private organisations, the largest being Demeter International, based in Germany. There are national Demeter associations in several countries. Organic certification of the vineyard is a *sine qua non* of membership. Demeter certification covers all biodynamic agricultural produce, not just wine. Biodyvin, on the other hand, is a French association for biodynamic wine-growers only. It has eighty or so members in France and one or two elsewhere.

## Common objectives

Common to these three groups – organic, sustainable and biodynamic – is the principle of being observant and proactive and ensuring that the vineyard is not hit, or at any rate is less badly hit, by pests. Insect pests are to be dealt with by augmenting microbiological life in the soil. The vineyard must be made easier for insects, birds and small animals to inhabit. The purpose of reinstating life in the vineyard is for Nature to take care of the balance between insect pests and their natural enemies. Prevention of fungal diseases, for example, demands greater presence and more manual labour in the vineyard. Conventional chemical products are to be replaced entirely (organic growers) or partly (sustainable growers) by natural preparations and ecologically approved products.

*Domaine Jean-Marc Brocard, Chablis.*

*Synthetic chemical pesticides are produced artificially.*

*Natural chemical substances, for instance, sulphur and copper, are also used in wine-growing.*

## The big clean-up

*Between 1993 and 2009 the European Commission carried out a major review of pesticides on the market. Something like a thousand different substances were affected and were evaluated in terms of their effects on human health and the environment. In the end, two-thirds of them were excluded from the market. It should be added, though, that many of them were removed because nobody applied to keep them. The products judged suitable (some 250 active substances) and new substances now approved by the EU have to meet the following requirements:*

- they have no harmful effects on consumers, farmers and local residents and passers-by;
- they do not cause unacceptable effects on the environment;
- they are sufficiently effective against pests.

*Eighty new substances have been added since 1989.*

# CHAPTER 3

# History in Brief

Within a very short time (in terms of world history), a farm using synthetic chemical products on crops and in the soil has become the norm. It is extremely uncommon nowadays for farming to be unaided by any kind of chemical pesticide of synthetic origin. Organic farming was fairly inconspicuous until just a few years ago. Now we have an explosion of newly converted farmers, wine-growers not least. Some people are actually wondering whether the tables won't be turned before long, with organic farming outstripping the conventional kind. Can we revert to pre-agrochemical times, and is it desirable to do so?

Now, we have to remember that pesticide use is older than the agrochemical corporations. There have always been pests and diseases to fight off. Sulphur has doubtless been used for two thousand years. Agents based on lead, arsenic, mercury, copper and nicotine sulphate were developed in the seventeenth and eighteenth centuries. A solution of lead and arsenic (*arséniate de plomb* in French) was copiously used as an insecticide in vineyards during the twentieth century, until it was found to be carcinogenic. It was banned at the beginning of the 1970s. The high lead content of the soils in Bordeaux, for example, may be due to lead arsenate. Pyrethrum and rotenone – today controversial 'organic' preparations – already began to be used as insecticides in the nineteenth century. So not all the products of yesteryear were healthy ones.

During the second half of the nineteenth century, the German chemist Justus von Leibig researched a way of adding nitrogen to the soil through artificial fertilisation. The practice became increasingly common after World War 1, following the development of the Haber-

*Poppies and vines at Domaine La Rune, Corbières, Languedoc.*

Bosch method, which converts nitrogen into ammonium, thereby making it accessible to plants, the point being that plants cannot absorb nitrogen directly from the atmosphere. The technology of artificial fertilisation came at a time of rapid population growth, and the higher agricultural yields it made possible were welcomed with open arms.

## Crisis in the vineyard

The vineyards of Europe were hit by several grave crises during the second half of the nineteenth century. Phylloxera, of course, was the worst of them, but two other serious diseases, also originating in America, now made their first appearance in Europe, namely powdery mildew (*oidium*) and downy mildew (*mildiou*). Powdery mildew was first spotted in French vineyards in 1847, downy mildew in 1878, only a few years after the phylloxera aphid had taken its first hesitant steps on European soil. A cure would have to be quickly found for all three pests if the European wine industry was not to be laid waste completely. Sulphur became the solution for powdery mildew, and the Bordeaux mixture, with copper as its main ingredient, was introduced for dealing with downy mildew. In the twentieth century these products were supplemented with synthetic pesticides. Downy mildew and powdery mildew remain two of the wine world's biggest problems.

## The post-war years

Just when the phylloxera plague had been defeated by grafting the European vine onto American rootstocks, two world wars occurred in rapid succession. After the first one, the transformation of agriculture, wine-growing included, was an accomplished fact. Work was mechanised and tractors superseded draft animals. Natural manuring disappeared and was replaced by artificial fertilisation. Farming became more intensive and rational. Copses, trees and hedges were cut down to make big, easily manageable fields with no obstacles in the way of

machinery. Weeds began to be eliminated chemically. The chemical companies, which during the wars had been busy researching chemical weapons, now discovered a peacetime use for their knowhow.

Perhaps the biggest change compared with pre-1914 was the transition from polyculture to monoculture. This one-sided farming affected the balance of nature both in the soil and above it. The natural enemies of insect pests vanished, and so chemical products were developed for dealing with pest and diseases. There was little concern over damage to the environment, at least to begin with. But one problem which perhaps had not been foreseen was that after a time insect pests became resistant to a certain product and a new and stronger one had to be developed. It was a vicious circle.

Weeding was real drudgery before the advent of chemical weed killers. Now weeds could be dealt with quickly and conveniently, but some farmers gained the impression that microbiological life in the soil was diminishing and, as a result, the soil itself was turning hard and compact.

A reaction against this new farming already came in the 1920s, in France, with the first organic movement. The members were mostly agronomists and growers who 'wanted to restore the balance of nature' and advocated roughly the same thing as organic farmers today: respect for the balance of the soil. Rudolf Steiner, whose teachings form the basis of biodynamic farming, lamented in 1924 that the earth was dying. But neither he nor the organic pioneers had much to show for their efforts. Artificial fertiliser gave the soil its principal nutrients – nitrogen, phosphorus and potassium – it was convenient to use, and chemical pesticides went on developing. Several more decades were to pass before organic farming gathered speed in earnest.

The organic movement had numerous firebrands in the twentieth century. In Britain during the 1930s, the agronomist Sir Albert Howard proclaimed the importance of restoring the fertility of the soil. The Swiss Hans Müller launched his ideas on organic farming at about the same time. In 1948 in France, Henri Charles Geoffroy founded the La Vie Claire co-operative, which now has two hundred organic food stores all over the country. Henri Charles Geoffroy was a rural romanticist to whom cities were the root of all evil. If only everyone would move back into the countryside, farming would recover its health. Other pioneers were blessed with business acumen. In France, for instance, Raoul Lemaire and Jean Boucher developed a method of fertilising the soil with calcified seaweed which they marketed through their Lemaire-Boucher company.

## Awareness

Things began to speed up a little in the 1960s, following the exposure of health and pollution hazards associated with a number of synthetic products (DDT not least). The organic movement took an upward turn. The French Nature et Progrès (N & P) organisation was formed in 1964 and in 1972 issued the very first organic standard. Today N & P's rules on organic growing are among the world's most stringent.

The fuel crisis of the 1970s raised awareness of the earth's resources not being unlimited. IFOAM (the International Federation of Organic Agriculture Movements) was formed in 1972. This is an umbrella organisation aimed at guiding and assisting the organic movement worldwide and laying down guidelines for organic growing.

In 1991 the EU issued its first Eco Regulation. Organic farming was now official.

## Organic wine-growing

Organic wine-growing began to gather speed above all from the 1980s onwards. Previous to that it had led a fairly discreet existence. Many people, presumably, had their first intimations of organic wine-growing through some of the early biodynamic growers who entered the scene in the 1980s and attracted a good deal of press coverage. Nicolas Joly converted his Domaine de la Coulée de Serrant, in the Loire Valley, to biodynamic farming in 1984. François Boucher, also located in the Loire Valley, already began farming the family estate, Domaine du Château Gaillard, biodynamically in the 1960s, later

*A flowering vineyard in Languedoc.*

becoming France's best-known biodynamic consultant. By the end of the 1990s, Domaine Huet in Vouvray, Marc Kreydenweiss in Alsace, Domaine Leroy and Domaine Leflaive in Burgundy, Pierre Frick in

*Bicycles are a common form of transport at Cono Sur in Chile.*

Alsace and Chapoutier in the Rhône Valley had also joined the ranks of the biodynamic *viticulteurs*.

It was the biodynamic growers who scored most headlines, presumably because they were well-known, respected producers already and Nicolas Joly was an eloquent, vociferous foreground figure (with a talent for saying things which were the stuff that banner headlines are made of). No wine connoisseurs doubted the high quality of these wines, though the majority, perhaps, did not really grasp what biodynamics was all about. The organic growers, unlike their biodynamic colleagues, did not have a famous name to help them look respectable. For a long time consumers doubted their ability to produce quality wines, and initially they were lumbered with a bad reputation.

Now their status has risen and the only question is where it will end. How many organic wineries will we have in ten years' time?

*Overleaf: Some wineries, like biodynamic Colombaia in Tuscany, leave alternate rows unploughed.*

# CHAPTER 4

# How Widespread is Organic Wine-Growing?

There is so much talk of environmental issues in the vineyard nowadays, not least in wine journals, that one can readily believe organic growing to be far more widespread than it really is. As yet, the acreage in Europe, and worldwide, with organic certification (that is, officially controlled and inspected) is very small.

The world today (2012) has about 260,000 hectares of organic wine-growing land, up from 160,000 in 2010. A drop in the ocean, considering that there are something like seven billion hectares of vineyards altogether – roughly 3.2 per cent, in other words. But more than twice the figure ten years ago. This is only the beginning.

The biggest organic acreages are in Europe, in the three big wine-producing countries: France, Spain and Italy.

Just over six per cent of all vineyard acreage in Europe today is organically farmed. Nearly all over Europe, the number of estates beginning to convert is increasing by the month. Talking to small and medium producers in, say, the south of France, one easily gets the impression that within a ten-year period everyone will have gone organic. Many growers actually believe that consumers are going to insist on it where exclusive, rather more expensive wines are concerned. The possibility cannot be discounted, but perhaps a little more than ten years will be needed.

## Spain and Italy top of the list

In 2010 Spain leapt to first place on the world list, with nearly 57,000 hectares of vineyard either certified as organic or undergoing

conversion. In 2012 it reached 81,000 hectares. Spain has the world's largest wine-growing acreage (roughly a million hectares), so perhaps this figure is not so remarkable, but it is interesting all the same, considering that in 2008 there were only 30,000 hectares with certification. Half of Spain's organic acreage is in Castilla-La Mancha, the plainland south of Madrid, where summertime temperatures are very high. Most of the organically produced wine from here is plain bulk wine for export to northern Europe.

Italy in 2010 had 52,000 hectares of organic vineyard, which is 6.3 per cent of the total. In 2012 it reached 57,000 and 6.8 per cent. The biggest organic regions in Italy are Sicily, Apulia, Tuscany, Marche, the Abruzzi and Emilia-Romagna. A large proportion of organically produced wines are exported.

## France

France's organic wine-producing acreage has gone from 5,000 to 65,000 hectares in a little more than fifteen years. Between 2008 and 2009 it increased by 39 per cent, and between 2009 and 2010 by 28 per cent. Today (2012) 8 per cent of French wine-growing acreage is organic (up from 6% two years earlier). If things go on like this, the organic acreage will soon be 10 per cent of the total, according to Agence Bio, the government agency responsible for the registration of organic farms and estates. In 2012 the number of organic vineyards in

France (with certification or undergoing conversion) stood at 5000, up from 3,945 two years earlier.

The three biggest organic regions in France in 2010 were Languedoc-Roussillon (16,462 organic hectares), Provence (11,209 hectares) and Bordeaux (7,715 hectares).

All regions of France are steadily increasing their organic acreage. Between 2008 and 2009 the biggest increase was in the southern regions, exceeding 50 per cent in Languedoc and Rhône. Between 2009 and 2010 it was Bordeaux (where many are actually convinced of organic growing being ruled out by the humid oceanic climate) and Champagne which topped the list, with an increase of 41 per cent each.

## Austria and Germany

Austria is the European country with the largest organic vineyard acreage in relation to its size, namely 10.2 per cent, up from 7.3 per cent two years earlier. In Germany too, the organic trend has momentum. Organic agriculture has a good reputation in Germany and is associated by consumers with high quality. Germany has always been ahead of many other European countries where acceptance of organic wines and food products is concerned. Over 7,400 hectares (up from 5,000 hectares two years earlier) of the country's total wine-producing acreage of 100,000 hectares is now organic or in the process of conversion.

## The New World

The New World lags behind Europe as regards conversion to organic growing. It is hard to say why. Home demand for organically grown grapes, at all events, has not yet gathered speed in the same way as in Europe (and Canada), added to which, the estates, specially in South America, are nearly always much bigger than in Europe, which is perhaps an impediment. Another impediment may be that the USA is alone among the new wine-producing countries in having official regulations. But people are now beginning to understand that organic farming can be a good selling point in export markets, so there are bound to be rapid changes imminent in these countries.

In the USA, 2.5 per cent of acreage is certified, in Australia 2 per cent, and in South Africa less than 1 per cent. Chile has 4,000 certified hectares, which is 3.6 per cent of total acreage, 110,000 hectares. In Argentina at present, 2 per cent of the acreage is certified.

## Organic wine-growing worldwide (figures for 2012)

| Country | Approx. total acreage (hectares) | Approx. organic acreage (hectares) | % of acreage |
|---|---|---|---|
| **Europe:** | | | |
| Spain | 1,100,000 | 81,000 | 7.4 |
| Italy | 818,000 | 57,000 | 7 |
| France | 780,000 | 65,000 | 8 |
| Germany | 102,000 | 7,400 | 7 |
| Austria | 48,000 | 4,300 | 9 |
| Switzerland | 15,000 | 270 | 1.8 |
| **New World:** | | | |
| California | 398,000 | 7,000 | 2.5 |
| Argentina | 173,000 | 4,000 | 2.3 |
| Chile | 200,000 | 4,500 | 3 |
| Australia | 160,000 | 3,200 | 2.25 |
| South Africa | 132,000 | 1,000 | 1 |
| New Zealand | 33,400 | 800 | 2 |

*Source: France Agrimer and Agence Bio.*

## Organic growing in France

| Region | Acreage with organic certification, 2012 (hectares) | Total acreage (approx.) | % organic of total acreage |
|---|---|---|---|
| Languedoc-Roussillon | 20,800 | 201,500 | 10 |
| Provence | 15,000 | 90,000 | 16 |
| Bordeaux | 9,700 | 130,000 | 7.4 |
| Rhône | 5,000 | 60,000 | 8.3 |
| Burgundy (+ Jura) | 2,600 | 50,000 | 5.2 |
| Alsace | 2,200 | 15,000 | 14.7 |
| Loire | 2,300 | 60,000 | 4 |
| Champagne | 400 | 32,000 | 1.25 |

*Source: Agence Bio*

*Overleaf: The view from My Wyn, Franschhoek, South Africa.*

# CHAPTER 5

# Who Goes Organic?

Organic growing comes easier in some places than others. It is of course easier in a dry, warm climate, as in France, Argentina or Chile, than in a cool and rainy one posing a greater risk of fungal diseases.

Organic growing is also possible in that sort of climate – plenty of examples can be quoted – but it requires a little more persistence. There are many organic growers in the Loire Valley, Burgundy and Germany, and numbers are steadily increasing in Champagne and Bordeaux.

Then again, not all organic vineyards are small ones, as one might be tempted to believe. The average size of a French organic vineyard in 2010 was 13.5 hectares (as against 12.0 in 2008), whereas the average size of vineyards generally is 8 hectares. The organic producers today number both big firms and co-operatives. Perhaps it is easier for a small family undertaking or a medium-sized estate to convert, but this could be because dedication and enthusiasm come more easily in a small organisation. A proprietor with many employees has to be able to motivate them, and many producers highlight this as the biggest hurdle.

On the other hand, having a large, contiguous estate can have its advantages. Contamination from neighbours' pesticides can be a problem in a district like Burgundy, where differently owned vines stand close together. Sometimes the organic grower tends the neighbour's nearest rows of vines to create a buffer against spraying. Another option is simply to sell off the grapes from the rows nearest the neighbour's.

*The Romanée-Conti Grand Cru vineyard in Burgundy.*

Certain organic methods of pest control – sexual confusion, for instance – will only work in a continuous area of more than five hectares.

The fact is that there are all kinds of organic vineyards, ranging from small ones producing exclusive quality wines to big ones of several hundred hectares going in for more modest wines.

## And why?

One sometimes hears people saying: 'I've heard that organic wines can be just as good', meaning: 'just as good as wines from grapes that have been sprayed with synthetic chemical pesticides.' Organic wines have always been saddled with a poor reputation. Could this be because organic vegetables looked so wretched and boring a few years ago? Or was it feared that wines made from grapes affected by rot and fungal diseases without conventional treatment being permissible would taste mouldy?

Whatever the reason, organic wines are now shedding their bad reputation, though it has yet to be entirely dispelled. Certain producers still prefer working ecologically without showing it, for fear of consumers considering organic wines inferior in quality or at any rate in no way superior. Opinions vary a great deal from one country to another. Consumers take a more positive attitude, for example, in Germany than they do in Italy.

As regards the taste of the wine, there are good organic growers just as there are bad ones, and there are organic wines which one likes and dislikes. In this respect, organic wines are no different from conventional ones.

## The environment matters most

So it is only quite recently that organic growing has begun to be thought of as a sales advantage. 'Today it's definitely a selling argument, above all in the export market,' says Remi Duchemin of Domaine Plan de l'Homme, Languedoc. Giovanni Menti, near Soave, agrees:

'We're selling our wines more easily, especially for export, now that we're organic.' But we have yet to hear any grower say that for him the financial arguments carry more weight than the ecological ones, though the former are doubtless at the back of people's minds. Many are doubtless being swept along at present by the general craze for organic produce. The more neighbours who go organic, the more difficult it becomes to stand aside. But most growers are genuinely convinced that this is good for the environment and that with organic grapes they will make better wines. Many say, like Emmanuel Pageot of Domaine Turner-Pageot in Languedoc: 'We want to be able to hand on a healthy winery to our children.' Christine Saurel of Domaine Montirius in the southern Rhône Valley concurs: 'When you have children running about in the vineyard, you'd rather not spray the vines with toxic chemicals.'

## Are organic wines more expensive?

Consumers are price-sensitive, and a producer will not automatically raise prices merely because the estate is now organic. But do organic wines cost more to produce? This is no easy question. You save on pesticides, but on the other hand more work is needed in the vineyard, and you may even have to take on an additional worker. The yield will presumably decline, initially at least, though with many organic growers it is back to the same level again after a couple of years. Then one must be prepared for the risk of losing parts of the harvest in certain years. There is no straight answer to the question of whether organic growing comes more expensive.

Many (but by no means all) producers say that it does – it is more labour-consuming, yields are lower, you have initial expenses which are only partly covered by the start-up grant of €350 per hectare obtainable from the EU.

Emmanuel Pageot of Domaine Turner-Pageot in Languedoc maintains that you get higher labour costs and roughly 20 per cent less yield, though, he continues, this is less of a problem for producers going for quality wines, because they do not aim for higher yields anyway. But, he says, you need to be able to position your wine in the market in such a way as to recoup the additional outlay.

Then again, comparing wine prices isn't easy. It is hard to tell whether it is the organic management of the winery that makes one particular Chianti or Languedoc wine cost more than another. Comparing the price of a kilo of tomatoes is much easier.

In 2011 Inter-Rhône, the growers' association, carried out an interesting experiment in the southern Rhône Valley with AOC Côtes-du-Rhône wines that were sold in bulk. When the 2010 vintage was offered for sale, it had been specified, for the first time ever, whether the wines were conventionally grown, organically grown or under conversion. The average price of a hectolitre of conventional Côtes-du-Rhône landed at €105 and that of a hectolitre of organically grown wine at €237. The conversion wines came in between. A considerable difference, and perhaps proof that the market is ready to pay more for wines made from organically grown grapes. At any rate, sometimes.

*The dramatic mountain formations in Franschhoek, South Africa*

# CHAPTER 6

# Organic Wine-Growing

*To convert = to change, to transform*

Going organic means a big change for the vineyard and the grower. One has to be aware of this and not be deterred by the first setback, be it an attack of downy mildew or grey rot. Converting a vineyard takes three years, and there is a reason for this. 'It needs time,' says Philippe Coston of Domaine Coston, Languedoc, 'and three years is not really long enough. Getting rid of all spray residues in the soil takes ten years or more.'

If you want to, you can go a little cautiously at first about reducing the synthetic products, so as to ease the transition, but from the first day of conversion the vineyard must be run 100 per cent organically. Which, according to the EU rules, means to say:

## *The following are prohibited:*
- artificial (synthetic) fertilisers
- chemical weed killers (herbicides)
- synthetic chemical insecticides and pesticides
- genetically modified products.

## *The following are permitted:*
- organic fertilisation
- spraying with copper
- spraying with sulphur
- certain products which are considered natural and are made, for example, from plants or micro organisms.

*Vines uprooted. Domaine de Cabasse, Rhône.*

Private organic labelling systems can have stricter rules than the EU's, for instance, concerning the amount of copper permissible.

Organic wine production in the EU, like all organic farming within the Community, is governed by Community regulation EU 834/2007. This stipulates which products may be used for spraying, fertilising, soil improvement, and so on. Only the products explicitly named in the text are permitted. Everything else is banned. Compliance with the rules by the certified farmers/wine-growers is monitored by private organisations which are approved as control agencies by the EU country concerned. Only those wine producers who have opted for certification, and thus consent to be inspected, are entitled to state on their labels that their grapes are grown in accordance with 'organic farming'.

As from the 2012 vintage, wine-growers within the EU are entitled to call their wines 'organic wine' and not just 'wine made from organically grown grapes'.

This is because in February 2012 the EU adopted rules on working procedures in the cellar and on the additives and production techniques permissible for an organic wine.

So, as from the 2012 vintage, we have the designation 'organic wine' on labels. The designation can also be applied retroactively to bottles from the 2011 vintage, if it is certified that the wine has been made in accordance with the new rules.

According to the EU definition of organic farming, however, a sense of nature should already have pervaded the *cave*:

> Organic Farming is an overall system of farm management and food production that combines best environmental practices, a high level of biodiversity, the preservation of natural resources, the application of high animal welfare standards and a production method in line with the preference of certain consumers for goods produced using natural substances and processes.

*Poppies on the Hermitage hillside.*

## How does organic differ from conventional?

Organic wine-growers use preventive measures as their main weapon against pests. In other words, they have to avert problems by being observant and vigilant, by tending their vineyard in such a way that it can withstand attacks. Otherwise they have only non-systemic (contact) products to spray with, for example, sulphur and copper. These are effective for a limited time only and are washed away by rain, added to which they are only preventive, which means that they have to be applied before the vine can come under attack. So an organic grower has to spend more time in the vineyard than conventional growers, who can spray their vines with systemic, penetrative products of more long-lasting efficacy.

Ideally, though, the organic approach should result in fewer problems, since the whole idea of organic growing is to restore the vineyard's equilibrium.

An abundance of life in the vineyard, with a diversity of plant species above ground and small bugs and insects below ground, is meant to stop diseases occurring. Pests must be kept down naturally by their enemies. That may be so in the best of all worlds, but it is not the case in the majority of vineyards. A modern vineyard is a monoculture,

which is an unbalanced type of agriculture, and so vines are in danger of both diseases and pests. How hard the vineyard is actually hit will depend on climate, weather conditions, grape variety, and so on. Sometimes delicate grape varieties are grown in damp climates which make diseases inevitable, even in organic vineyards. But an organic producer who goes about things the right way can at least mitigate the nastiness.

There is little to be done about the fact of wine-growing today being a monoculture, but one can improve the situation by planting hedges, flowers, trees and other plants round the vineyard, so as to create a more inviting habitat for birds, animals and insects.

## Good soil

Soil analyses carried out in the 1980s, for instance by the microbiologist Claude Bourguignon, revealed soil impoverishment in various wine-growing regions of France, his native Burgundy not least. Big newspaper headlines ensued on the theme of 'effectively dead soils'. In addition, years of chemical pesticides and heavy-tractor use had rendered the soil compact and impenetrable. A soil, Bourguignon insists, has to be

airy and porous, 'like couscous. Preferably you should be able to stick your whole hand into it, that's how friable it should be.'

An airy soil has the advantage of drying faster and, accordingly, warming up faster. Whether or not this imparts more of a *terroir* flavour to the wine is a moot point, but the vines will at any rate be less prone to water stress during hot periods.

## Roots matter

It is mostly through its roots that the vine absorbs the nourishment it needs from the soil. The further down the roots extend, the more able they are to supply the leaves with the nourishment the vine is in need of, among other things in order for photosynthesis to work properly. This in turn reduces the risk of drought stress and nutrient deficiencies.

It is important, then, to facilitate the spreading of roots from the moment of planting. Irrigation should only be resorted to if absolutely necessary. Perhaps the young plant will need a little watering so as not to suffer from water stress, which will affect the whole of its future development. Some growers advocate avoiding the plastic tubes (often blue in colour) which are placed round the young vine to fend off rabbits, on the grounds that they create moist, warm conditions that encourage the roots to stay on the surface.

Every type of rock contributes its minerals to the wine – zinc, copper, manganese and so on. Scientists have yet to make absolutely clear how this is revealed in the wine's aromas, but it has been observed that vines and other plants fertilised solely with NPK (see below) contain fewer trace elements, which may possibly give the wine less character.

*A Médoc vineyard with neither grass nor other plants.*

# What does the vine need?

## *Fertiliser*

The organic wine-grower has to help the soil provide the vine with exactly what it needs in the form of nourishment, minerals and water. A vine intended to have a small yield for the production of quality wine seldom needs to be given much fertiliser or nitrogen, but when the grapes begin growing after blooming, the soil may have difficulty, unaided, in supplying the vines with all the nitrogen they need.

Both nitrogen deficiency and excess of nitrogen cause problems. Nitrogen deficiency will cause poor growth, and the vine will sprout small leaves that fall off prematurely. Excess of nitrogen causes excessive vine growth, rendering the vine more vulnerable to fungal attacks, added to which, the grapes will be oversized and their juice diluted, resulting in a wine which is also diluted. The taste of the wine can be affected in both instances.

In addition to nitrogen, there are other nutrients which vines need to absorb from the soil. Phosphorus and potassium are the most important, and small amounts of trace elements such as iron, magnesium, zinc, copper and chlorine may also be needed. Artificial fertiliser was created for efficiently adding nitrogen, phosphorus and potassium to the soil, and is often called NPK, after the chemical designations in the periodic table. Organic growers have to use organic fertilisation instead, so as to give the soil the nourishment it requires.

The EU rules carefully specify what is permissible for fertilising and soil improvement. In order to give the soil what it needs, the organic grower may administer:

- animal and vegetable composts
- animal dung
- guano (high-phosphorus sea-bird droppings)
- algae
- natural calcium carbonate (calcite) and magnesium carbonate (magnesite)
- natural potassium sulphate, calcium sulphate and magnesium sulphate

*Weedazol weed killer.*

- 🌿 wine lees
- 🌿 composted grape skin residues
- 🌿 trace elements
- 🌿 biodynamic preparations

Soil that vines grow in often has low humus content, and sometimes offsetting the annual loss resulting, for instance from soil erosion, is not enough, and more has to be added in order to attain good balance. Organic fertilisation makes up for the lost humus. In addition to fertiliser, the humus is supplemented with ground branches from pruning, dead leaves and ploughed-in vineyard grass.

Organic fertiliser is spread over the vineyard in winter or early spring, so that the nitrogen will have time to mineralise (turn into nitrate) and can be absorbed by the vines in early summer.

69

## Artificial Fertiliser

*The difference between artificial and organic fertilisation is that artificial fertiliser, often administered in the form of nitrate, is highly soluble and is rapidly absorbed by the plants. Environmentalists argue that there is more risk of spillage with artificial fertiliser, because not all the nitrogen is absorbed by the plants. Instead, some of it leaches out into rivers and groundwater and can cause eutrophication and algal bloom. Organic fertiliser, it is true, also harms watercourses, but, the environmentalists maintain, to a lesser degree.*

*What is more, artificial fertiliser adds only a few substances to the soil, with the result that, sooner or later, the composition of the soil will change. Nitrogen, phosphorus and potassium are important substances, but there are more elements which the vine also needs.*

*Production of artificial fertiliser requires heavy inputs of energy.*

## Nitrogen fixers

If nitrogen is what the soil needs, this can be administered using other methods with less environmental impact.

Nitrogen is present in the atmosphere, but the plant must be helped to absorb it. One way of doing this is by means of a nitrogen fixer, such as leguminous plants. France in 1950 had three million hectares planted with the tiny flower called lucerne. Today there are only 200,000. Lucerne, clover and other leguminous plants have the advantage of being able to convert atmospheric nitrogen into nitrogen nutrient which other plants can absorb. So planting a leguminous species in a vineyard is a supremely ecological means of laying on a supply of nitrogen. Artificial fertiliser began putting leguminous plants out of business after WW1, but now they are coming back into their own.

## Phosphorus and potassium

Phosphorus is needed in order for photosynthesis to work properly. Phosphorus is present in the soil (more abundantly in some regions than in others) and is absorbed by plants. Copious fungal mycelium can facilitate phosphorus uptake. Phosphorus can also be added in the form of ground quartz. Biodynamic growers also use valerian, a plant rich in phosphorus, as a natural source of supply.

Potassium is the third vitally important element for the vine. It too is present, to a greater or lesser extent, in the soil. Potassium deficiency causes the leaves of the vine to turn pale and eventually die. Too much potassium in the soil gives a wine with high pH, which is bad for colour stability.

## Compost

Compost is important for organic farming and still more so for biodynamic farming.

Composting is a controlled degradation of organic and degradable waste. The waste itself may come from the plant or animal kingdom. Heat is generated during composting, and the compost loses volume as a result of water evaporating. This leaves a homogenous structure which is not malodorous and has an augmented humus content. The waste, one can say, undergoes a kind of 'purification'.

Compost gives more humus than uncomposted fertilising. Animal dung used in organic farming has to be composted if it comes from non-organic or industrial stock farming. The organic grower maintains that all animal excrement should be composted, because nitrogen from uncomposted manure is more soluble. Producing one's own compost is the best of all options, the advantage being that one can make it exactly as one wants to, with a content providing the exact nourishment that one's own land is in need of. The drawback is that producing one's own compost is a time-consuming business. For example, the compost has to be stirred, turned and watered regularly. Industrial compost often contains too much nitrogen in relation to humus. Then again, if you make your own you can spread it at various degrees of composting.

Compost adds potassium, phosphorus, calcium and other minerals to the soil. What it contains, and how much, will depend partly on its original constituents.

## Weeds and cover crops

Organic growers may not employ chemical weed control. They can only remove grass and weeds from the vineyard mechanically. But the best way, many maintain, is not to systematically remove all grass but to let the grass/weeds grow or else plant a cover crop. This has so many advantages that it has now become common practice, and not only with organic growers. Aside from the avoidance of using a chemical product, there are five main reasons for letting the grass grow:

- it reduces soil erosion.
- it improves the soil structure, thereby reducing the risk of compaction.
- it promotes microbiological life in the soil.
- it competes with the vines for water and nourishment, thus limiting the yield.
- it makes it far easier to move around in the vineyard, on foot or with a tractor, after a fall of rain.

One can let the grass grow freely, but this does not suit everyone. Alternatively, one can plant a cover crop, for instance, clover (the nitrogen fixer) or some type of grass or flower that will suppress the weeds. Whatever the type of grass or cover crop used, it will make a generous humus contribution when ploughed into the soil.

Some studies suggest that a vineyard with a cover crop develops more fungal mycelium than other vineyards. Mycelium is a type of fungus which lives in symbiosis with the roots of the vine and improves the efficiency of its water and mineral uptake, making it easier, for example, for the roots to absorb phosphorus, zinc and copper.

Having grass or other plants in the vineyard, however, does not always suit all vineyards or all wine-growers.

Certain soils/climates/grape varieties cannot cope with too much competition from grass for water and nourishment, especially during a hot, dry summer. This makes it important to choose the right cover crop, one that will compete with the vines as little as possible. One may

*A vineyard in Médoc, with the grass under the vines removed by means of chemical herbicide.*

choose to plough the grass in during summer or remove it from every other row, so as to reduce the competition. If the vine suffers from nitrogen deficiency owing to competition from grass, this can impart a reductive taste to the wine. Nitrogen deficiency can also affect the aromas in other ways.

The best solution of all is if one can find a grass to plant which will not disturb the vine and which therefore need not be removed so often, for driving a tractor day in and day out is not the most organic of practices.

Claude Bourguignon recommends ploughing as little as possible. It is, he says, not ploughing that keeps the soil aerated but the tiny creatures below the ground surface, and ploughing disturbs them.

In addition, heavy tractors tend to make the soil hard and compact, as well as – not least – entailing carbon dioxide emissions. Ploughing also has the effect of releasing carbon dioxide from the soil.

Ploughing away the grass in the vineyard is far more time-consuming than the use of herbicides, especially as regards removing the grass from right under the vines. Avoiding damage to the vines is a very tricky business, and the tractor has to be driven very slowly. This is also more expensive, because it means more labour input. There are many (non-organic) growers who see no other solution than using chemical herbicides to eliminate vegetation directly under the vine.

## The comeback of the hedge

An organic grower, then, wants a balance between the soil and life in the surroundings of the vineyard, with insect and plant life flourishing. One means to this end is to let the grass grow in the vineyard or to sow grass or some other cover crop. Clover, as we have seen, is good because it facilitates nitrogen uptake. Hedge-planting in vineyards is also gaining popularity. There are few organic (or for that matter non-organic) wine-growers nowadays who do not mention the word 'hedge' sooner or later in the course of conversation. Many metres of hedge are now being planted. Perhaps the French growers are aiming

*A vineyard nesting box to encourage birdlife. Le Fraghe, Bardolino.*

to replace the 200,000 km of hedgerows that were removed in France during the 1960s to make room for the new, highly mechanised agriculture.

At Château Guiraud, Sauternes, they are very proud of their achievement. Not only has this big (100-hectare) estate recently (in 2011) obtained organic certification, but they have now also planted six kilometres of hedges. But why this thing about hedges?

There are several reasons, one of the most obvious being that hedges serve as windbreaks, added to which, the fact is that they even out temperature variations in the vineyard. In autumn the hedge imparts natural fertiliser to the soil by shedding its leaves. Another important point is that the hedge becomes the habitat of useful insects

*Insect hotel at Château Guiraud, Sauternes.*

and birds, affording them food and shelter. Several new plants and insects have popped up at Château Guiraud, including a number of long-vanished insect species.

To offset the monoculture which wine-growing has now become, you need surrounding areas with different vegetation, trees, flowers, meadowland and so on. Simple measures like putting up nesting boxes for a small outlay will help to augment birdlife, and the boxes can be designed to attract precisely the bird species one wants – tits, owls, and so on – because they will help to keep down insect pests. Bats too are helpful in this respect. At Château Guraud they even have an insect hotel!

# CHAPTER 7

# Pests and Diseases

Even with flora and fauna in position in the vineyard and with the soil flourishing, things may still happen which have to be dealt with. The weather can bring fungal attack such as downy mildew, powdery mildew and botrytis (grey rot).

Insect pests, both new and old, can strike a vineyard regardless of its equilibrium. Few organic wine producers can get by without spraying once or twice a year. A limited amount of copper and any amount of sulphur may be used, as well as various products deemed 'natural' and harmless to animal and plant life and human beings.

## Fungal diseases

Just as with conventional wine-growing, fungal diseases (downy mildew, botrytis and powdery mildew) are what the organic wine-grower fears most.

### Downy mildew

Downy mildew (*mildiou* in French) is a disease which came to Europe from America and was first discovered in France in 1878. It is caused by a fungus, plasmopara viticola, which attacks the twigs or leaves of the vine or the grapes themselves. This disease is also called peronospera.

It can hit the vineyard at any time of the growing season, during cool, rainy weather. Yellowish patches appear on the leaves. The disease delays ripening and makes the plant more vulnerable. The

*A leaf with downy mildew.*

first treatments take place when the first affected leaves are spotted, and they then continue, possibly for the rest of the growth season, depending on the weather situation.

By tending the vineyard continuously, the wine-grower can delay the occurrence of downy mildew attacks or reduce their intensity. The grower should maintain a low yield (many organic growers say this is the very best way of avoiding disease problems), choose a suitable cover crop, rootstock and grape, and make sure the foliage dries quickly after rain and damp by removing leaves, tying up branches and so on. It is important not to plough the soil during critical periods, otherwise infection from the soil may spatter up onto the leaves.

The only really effective antidote to downy mildew available to the organic grower is copper, in the form of copper sulphate (commonest), copper hydroxide, copper oxide chloride or copper oxide. Most often the vineyard is sprayed with copper in the form of a solution known as Bordeaux mixture *(bouillie bordelaise)* and consisting of copper sulphate (so the solution also includes sulphur), slaked lime and water. It is pale blue in colour. Often it is purchased in powder form and dissolved in water. This is why one can see vines with blue leaves in summertime.

## Powdery mildew

Powdery mildew *(uncinula necator* in Latin, *oidium* in French) affects the vineyard in hot, humid weather. Attacks in France and elsewhere have grown more frequent in recent years. Preventive methods against them are less effective than those used for combating downy mildew, but are not to be ignored for that. It is important to keep the foliage aerated, because then it will dry better. The organic grower treats the condition with sulphur, the amount of which has not yet been regulated by law. It should be limited as much as possible, however, because sulphur is not harmless. Typhlodromus species, for example, do not like sulphur. These are helpful mites which live on the vine leaves and eat up other mites which are dangerous to the vine.

Copper also has a secondary effect against powdery mildew.

Alternative products are being tested. The results vary a great deal and, unfortunately, are none too encouraging on the whole. One alternative product is fenugreek, a so-called SDN *(Stimulateur de Défense Naturelle)*, a plant stimulator classed as a natural pesticide. SDNs are substances which can trigger the plant's own defence system once it is attacked by a disease (rather like vaccination). A substance of this kind which organic growers are allowed to use for combating powdery mildew is derived from fenugreek.

## Botrytis

There is no really efficacious pesticide for botrytis (grey rot) which organic growers are permitted to use. Instead they use the same preventive agents as for downy mildew, and several natural products are currently being tested. One of them is ground limestone, which has a desiccating, healing effect, and another is citrus extract. In addition, copper also has a secondary effect against botrytis. A fungus called trichoderma has proved effective. This is sprayed on the vines as a cure. It is used, for example, by organic wine producers in Chile and Australia and is being experimentally used in France.

## Other Pests

## The Grape Worm

The grape worm *(ver de la grappe* in French) is in fact the larvae of two different moths: cochylis and eudemis. One has to be on one's guard against the grape worm throughout the growing season, because the butterflies lay eggs three times between March and August. The first-generation larvae devour the tiny buds, thereby reducing yield. The second and third generations crawl inside the grapes, which can mean a risk of lower yields but also of botrytis, especially in damp weather. The larvae spread toxic fungal spores which provide nourishment for botrytis.

In conventional wine-growing, the grape worm is killed with chemical insecticides. In organic growing there are various alternatives: sexual confusion, bacillus thuringiensis, trichogramma, spinosad and bats, for example.

Sexual confusion is based on spreading the scent of pheromones (a substance secreted by the female to attract the male) in the vineyard by depositing capsules of the hormone at strategic points. The male becomes confused and cannot find a female, and so no reproduction occurs. The capsules have to be hung up from the end of March, and for the method to be effective the vineyard must not be too small. If it is less than five hectares the method will not work, but one can join forces with the neighbours to increase the acreage. The distribution rate is 500 capsules per hectare.

This method is perfectly harmless to the environment, wildlife and humans.

Champagne is the region of France where grape-worm prevention through sexual confusion has advanced furthest. Nine thousand hectares – 25 per cent of the total acreage – are now protected in this way, and the plan is for the figure to rise still further. But this, of course, is not to say that 25 per cent of the Champenois acreage is organic! Sexual confusion is one of the environmentally friendly methods which non-organic growers have also resorted to.

The grape worm can also be combated with a bacterial extract called bacillus thuringiensis. The larva, that is, the grape worm, eats the extract, which contains lethal toxic substances. This is effective, especially against eudemis, less so against cochylis, and it is absolutely harmless to humans and the environment.

Experiments are also in progress using trichogramma, a tiny parasitic wasp which is introduced in the vineyard and kills the larvae while they are still at the egg stage. Scientists are cautiously optimistic, but have not yet been able to draw any firm conclusions.

Bats readily devour the cochylis and eudemis moths, and so it is a good idea to make them feel at home in the vineyard or its surroundings.

Spinosad is a new insecticide which was recently made permissible for organic farming within the EU. It is produced by fermenting a bacterium, Saccharopolyspora spinosa, which was first discovered in the soil in the Caribbean, underneath an old abandoned rum distillery (!). Spinosad acts both by contact and when eaten by the insect. It produces good results with high mortality (among the insect pests) and can

*A bunch of grapes affected by botrytis (grey rot).*

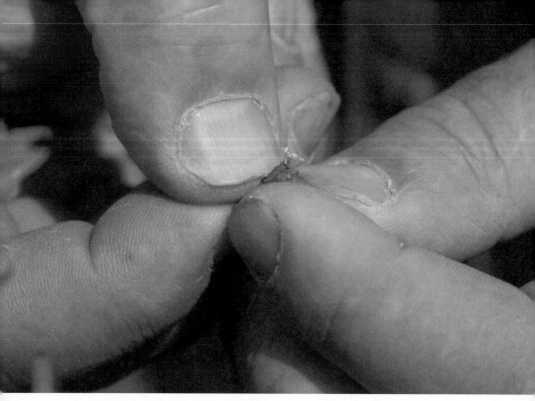

*This little grape worm can do great damage.*

be used, for example, against the grape worm. It is considered to do little harm to other insects and animals, but one must bear in mind that it is toxic to bees and must therefore be used with care.

## Red spider mites

The little red spider mite makes its home on the leaves of the vine and can do a lot of damage by chewing on them and thereby obstructing photosynthesis. These spiders breed rapidly, and when the number climbs to fifteen or twenty per leaf you have a problem. The leaves turn yellow and die. Typhlodromus mites, a predatory species, are their natural enemies.

Typhlodromus mites are present to a greater or lesser extent in all vineyards, and they seem to be more numerous in organic vineyards and vineyards with cover crops (but this point is still being researched).

*Sexual confusion via capsules is one way of combating the grape worm.*

## *Flavescence dorée*

*Flavescence dorée* is a serious bacterial disease which came to Europe from the USA in the 1950s. The rootstock is weakened and can even be killed by it. The disease can have a long incubation period. To keep it from spreading, one has to uproot the infected rootstocks and combat the grasshopper that spreads the disease. It is compulsory in France to report to the authorities if one has affected vines, and there are some regions where spraying to eliminate the grasshopper is obligatory.

For a long time, grasshopper control remained a problem to organic growers. Until very recently, rotenone was all that could be used. This is a substance obtained from certain tropical plants, and in addition to being toxic to several animal species, its use is believed to be connected with certain cases of Parkinson's disease. It was banned in 2009, but growers in several countries, France among them, were exempted till October 2011.

Organic growers have been able since 2009 to combat the grasshopper with a product called Pyrevert, based on pyrethrin, a substance derived from a chrysanthemum growing in Kenya and Australia. Pyrethrin is toxic to bees, Typhlodromus mites and certain other insects, and can be said to nudge the limit of proper approval for organic growing, but even so it is less toxic than rotenone and far more effective.

Sans sulfites

Sulfite-free

# CHAPTER 8

# Pest Control by Natural Means

## PNPP *(Préparations naturelles peu préoccupantes)*

Infusions and decoctions of plants and herbal teas have been put to agricultural use in all ages. For centuries past, knowledge has existed concerning the use of natural products from just round the corner for the avoidance and cure of diseases and damage affecting various crops. Products of this kind have now acquired a name of their own in France: PNPP *(préparations naturelles peu préoccupantes),* that is, 'natural preparations giving little cause for concern'.

The name became official in France in 2009. French law defines a PNPP as a product which is so easy to make that anyone can do it, is not toxic, does not endanger health or the environment and is not patented. You might think that an organic grower was quite free to use such products, that all one need do was pop round the corner and pick a few nettles and some horsetail. Well, it is hard to stop anyone from picking them but you are not allowed to manufacture and sell any product obtained, nor (by the letter of the law) to disseminate any recipes in books or on the web unless the product has gone through all the hoops and been approved for use and sale. All products, PNPPs included, have to be analysed, inspected and approved before they can be used for plant protection in farming.

## The nettle war

Understandably, many organic growers who are long-time users of nettles find this policy decision hard to swallow, not least the French grower Eric Petiot, who in 2006 had his computer impounded and was threatened with a heavy fine for publishing a book on how to make your own nettle water. The period of protests and conflicting arguments that followed was dubbed the Nettle War. After various ins and outs in parliament, the politicians promised to fast-track the issue, but it still took until 2011 for nettles to become the first PNPP approved for use in France.

The recipe – the only one now permissible if the nettle water is to be sold– was published in May 2011 in the *Journal Officiel,* as are all new laws and regulations. Many nettle-expert bloggers, however, objected that this was not the recipe commonly used, and so there seems to be a considerable risk of illegal nettle waters coming on to the French market. *Journal Officiel* also specifies that nettle water may be used against downy mildew and various mites, and also to stimulate plant growth.


## Approval of organic plant protection products

A product to be used for plant protection in organic farming within the EU must:

- be inscribed on the EU list of approved substances (EU Regulation 834/2007),
- be inscribed on the EU list of substances approved for use in organic farming (EU 889/2008),
- have been given what is termed Autorisation de Mise sur le Marché (AMM), that is, marketing authorisation (this applies in France).

Administration is burdensome, and certain substances are now included in one list but not in the other, in which case no AMM is obtainable. Getting a product registered (moreover, both in the EU and in France) is a very expensive proposition, costing between €10,000 and €100,000. Organic plant protection products have little commercial value and are often sold by small entrepreneurs, and so raising the money for getting permits can be difficult.

Is it right or wrong that preparations of this kind should be subject to the same control and approval procedure as synthetic chemical pesticides? In a way one can understand 'natural' products having to go through the control machinery if they contain an active substance which affects an agricultural product. This may seem a trifle exaggerated where nettles are concerned, considering that all you do is pick young nettles, put them to soak in water and then add more water. But on the other hand, a product is not always harmless just because it is natural. Pyrethrin, which we have already mentioned, is a case in point.

Research into natural products is only just beginning. The important thing now is to define how they work, whether they are toxic, whether they reinforce the plant's immune defence, and so on.

Gil Rivière-Wekstein, a journalist specialising in agricultural and environmental affairs, claims that organic plant protection products based on plants and vegetable matter are often the precursors of synthetic products. In 2011 he published, in France, a widely noted book entitled *Bio – Fausses promesses et vrai marketing* (Organic – False promises and true marketing), in which he heavily criticises organic farming. Synthetic products, he argues, have the advantage of acting longer, so that treatments do not have to be repeated so often, added to which, the active substances are kept under control. He goes on to say that the amount of active substance a natural product contains may depend on where it grows, the age of the plant and other things.

## Copper and its alternatives

The use of copper is the big issue where organic farming is concerned. Copper is a heavy metal and, in heavy concentrations, is toxic to soil and water.

Organic growers are often criticised for using more copper than their conventional colleagues, who combine copper spraying with synthetic products (to which the organic growers rejoin that copper, at any rate, presents no danger to the person who does the spraying). So long as there is no good alternative to copper, organic growers will have to put up with being criticised and instead keep their copper

spraying to a minimum, with the aim of reducing it to a level where it will not endanger the soil. A scientific study of 'the state of French soils' *(L'Etat des Sols de France),* conducted by GisSol on behalf of the French departments of agriculture and the environment and the INRA research agency, among others, was published at the end of 2011. It shows all French wine-growing districts to have high concentrations of copper in their soil, that is, over 50 mg per kg. The highest concentration recorded was 322 mg.

Ten years ago, a complete ban on the use of copper by organic growers began to be mooted within the EU, but, realising the impossibility of banning it altogether, they opted instead for restricted use. The first limit – 8 kg per hectare annually – was imposed in 2003, and in 2006 this was lowered to 6 kg.

At present, then, organic growers are allowed to use up to 6 kg copper per hectare each year, or, more correctly, 30 kg per 5-year period, which means that the amount used can vary from year to year. France's Agence Nationale de Sécurité Sanitaire de l'Alimentation, de l'Environnement et du Travail (ANSES, a government agency reporting to the ministers of health, agriculture and the environment) recommends lowering the maximum limit to 4 kg, but studies have shown this to be impossible. ITAB says that 4 kg will not afford adequate protection in years with heavy attacks of downy mildew.

Even so, the use of copper could be kept down if combined with proactive measures. The aim must be to make it hard for the fungus to establish itself in the vineyard. Pruning and aeration of foliage are important measures. By keeping the foliage well aired and ensuring that it does not become too dense, one increases the effect of dry winds. Swift response must be possible when weather and rain forecasts demand it. It is also important for the spraying machine to be properly adjusted, so that none of the spray will be wasted or land in the wrong place.

*Spraying with Bordeaux mixture turns the leaves blue.*

## Copper dosage

Research is underway to ascertain the best dosage procedure and the smallest doses which are really effective. Use can be reduced to 500 g copper per hectare and spraying operation, but research has shown the ideal (for efficacy and the environment) to be between 600 and 800 g. Of course, the less one sprays each time, the more precise one has to be (machine set properly, and so on), so that everything lands just where it should.

A dose of 300 g per hectare can be enough to deal with a mild attack of downy mildew. In the event of heavy attacks, especially early on in the season, in May or early June, using less than 600 or 800 g copper per treatment is risky.

In a worst-case scenario, a vineyard may need up to thirty treatments in a year, but this is unusual. The average for a difficult year is between nine and twelve treatments. In humid regions (Champagne, Burgundy, the Loire and Bordeaux) the average is between five and fifteen. Altogether between 4 and 6 kg copper will be used in a bad year. The amount used in a less difficult year varies between 2.5 and 3.6 kg, administered in 5 to 7 treatments.

Copper acts by contact, and so it remains on the leaf after treatment. Rain washes it off, and the treatment may have to be repeated – for the protection of new shoots – every ten days. Treatment followed by an unexpected fall of rain may have to be repeated sooner.

Conventional and sustainable growers can use systemic plant protection products instead. These are absorbed by the sap in the vine and protect the plant from inside. This is the most effective product from a protection viewpoint, because it can also protect new shoots. Alternatively, these growers can use penetrative products which enter the plant through the leaf stomata. These products do not protect new shoots.

## The alternatives to copper

Although efforts are now being made to find more environmentally friendly alternatives to copper, this is more a question of reducing the amount of copper than eliminating it entirely. Experiments, which have yet to be evaluated, have been in progress in Bordeaux and elsewhere since 2010, using infusions of horsetail, wormwood and peppermint which are mixed with a small amount of copper. New experiments with nettles and lactoserum are also in progress. Spraying copper together with trace elements such as zinc and boron may also be an interesting proposition, especially at the start of the season.

The possibility has also begun to be explored of using a plant stimulator *(Stimulateurs de Défense Naturelle, SDN)* together with smaller doses of copper. The SDNs in question include, for example, a microscopic fungus called trichoderma hardianum, an alga called laminaria and copper gluconate.

## Spraying

Proper care of the spraying machine is vital for limiting the amount of spraying. One has to ensure that the products land in the right place, that the machines are properly set, that there is no spillage, that none of the spray misses the target, and so on. Spraying operations are also required to show consideration for what the French call *ZNT – zone non traité*, that is, an untreated area. The minimum distance to be maintained between the crop treated and a watercourse is between 5 and 100 metres or more, depending on what product you are spraying with. There are also rules about at what times spraying is allowed. It is not permitted in rainy or very windy weather.

### *Watch the weather*

The severity of fungal diseases depends very much on the weather, which puts the weather and weather forecasts at the centre of wine-growers' attention, and with today's technology there is no excuse for not being prepared for whatever the elements may get up to.

How does the wine-grower keep himself updated on weather conditions? Does he take in the forecast at the end of the television news programme, like the rest of us? Perhaps, but the majority probably have more sophisticated methods up their sleeves. France has several on-line services for farmers, giving detailed weather information for different

*Bags and cartons of pesticides and fungicides.*

*A weather station.*

zones in a region. Growers can subscribe to services providing detailed information for their own districts. They study satellite images of cloud formation. Rain and other potentially troublesome phenomena, such as hail and frost, are often very local. Geographic precision is vital. And so wine-grower associations have begun setting up weather stations here and there in various districts. It is better still if the individual grower can afford to put up a weather station in his own vineyard (or several if he has a big estate).

The weather station has sensors which gather data for transmission to a computer. In this way the wine-grower can obtain continuous, reliable local information in real time concerning air temperature, air humidity, wind speed, solar intensity and the risk of precipitation. The more information one has about the weather, and the better that information is, the more efficiently one can go about things in the vineyard. Knowing how the wind is blowing, from what direction, and if there is going to be rain enables one to decide whether and how to spray. With dependable information about imminent rainfall, one can also reduce the number of

*Cleaning the spraying tractor.*

spraying operations. You will not spray, for instance, if you know it will soon be raining, because the rain just washes the active substance away, and then you will have to spray all over again when the weather clears.

Data from the weather station can also be used for judging how much moisture is evaporating from the soil and, accordingly, how much water the vines are in need of. This is particularly useful in arid regions where growers are dependent on irrigation. To be still more advanced, you can have buried sensors which measure the temperature and moisture content of the soil, so that you can tell more easily how much watering is needed. Sensors can also be installed inside the foliage. They then measure its dampness, which improves the chances of keeping fungal diseases and insect pests in check.

## Crosses and hybrids

There's no getting away from it – Vitis vinifera, the European family of grape vines, is a delicate plant. So what do you do if you are in a district prone to downy mildew and botrytis and don't want to drench your vines in spray mixture? One solution is planting hybrids, that is, crosses between a vitis vinifera and an American vine. The only problem is that many people in France and other established wine countries would turn up their noses at these environmentally friendly vines, which are deemed qualitatively inferior to pure vinifera vines.

If you plant the right sort of hybrid, you will get a plant with the American vine's resistance to various diseases (downy mildew included) and some of the qualitative properties of the vinifera. These modern hybrids are termed *hybride interspécifique* in French, to distinguish them from the crosses of vinifera and American vines which were made a hundred years ago to produce grapes resistant to the Phylloxera louse. The modern hybrids have been through several different crossings and may include up to 70 per cent vitis vinifera, in which case they have to be grafted onto American rootstocks like any other vitis vinifera.

These hybrids are often developed in countries like Germany and Switzerland, and instead of being resistant to the Phylloxera louse they are highly, sometimes very highly, resistant to fungal diseases such as powdery mildew, downy mildew and grey rot. Sometimes the grower does not even need to spray the vines with copper. So these grapes, without a doubt, are environmentally friendly, which is a point the growers using them – mainly British, Belgian and Swiss – are at pains to drive home. Philippe Grafé at Domaine du Chenoy, near Namur, describes his *pinotin,* a Swiss-made cross between pinot noir and an unknown grape variety. He says that he only needs to spray his vineyard three times a year, instead of the twelve to fourteen times which are common practice in northern European vineyards.

It remains to be seen whether the hybrids have a future in bigger wine regions. Today it seems unthinkable for a grower in Bordeaux, Champagne or Burgundy to start using a hybrid, but you never know. Perhaps hybrids can win acceptance more easily than vines which have been genetically modified to make them disease-resistant. In France at present, hybrids are only permitted to a very limited extent.

## Genetically modified grape vines

The organic rules prohibit all genetically modified (GM) products. Needless to say, this includes genetically modified vines, which are completely prohibited in the EU, along with other genetically modified products in wine-making – yeast, for example. As in so many other cases, however, this is a question to which there is no straight answer. The basic philosophy of organic wine production is care for the environment, and genetically modified products are not *ipso facto* environmentally harmful. Nor are genetically modified grape vines. R&D might possibly lead to GM vines which are immune from, or at least highly resistant to, various diseases, for instance, powdery mildew, downy mildew or *flavescence dorée*. This would, for example, eliminate the present need for spraying vineyards with heavy metals by way of disease prevention, which would in many ways be good for the environment. Others maintain that continuing research in this field would entail unacceptable risks because, the argument goes, the consequences are not wholly foreseeable, and so certain groups want to forestall any further activity in this research field.

But the whole debate on genetically modified grape vines and other GM products today is very political and polarised, and so we will not delve into the issue any further. All genetically modified products today are banned from organic growing, and things appear set to remain that way in the foreseeable future.

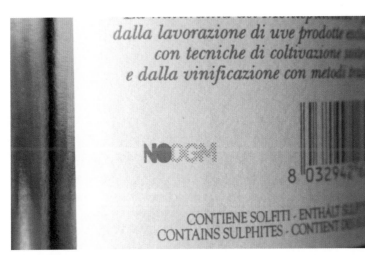

*'No OGM' – No genetically modified products.*

# CHAPTER 9

# I Want to go Organic – How is it Done?

Conventional practice today means spraying agricultural products and land with greater or lesser quantities of synthetic chemical pesticides and providing nourishment to the land with artificial fertiliser. If you go organic and, accordingly, dispense with these products, you have to obtain certification in order to state as much on your label.

Of course, there is nothing to stop anyone practising organic farming regardless, without certification, but in that case you will not be allowed to label your products (wines) as organic. It is hard to say how many growers choose to go without certification, and hard to say whether they really are organic. It is up to consumers to decide which growers they trust.

The trend today, though, is for more and more to obtain certification. Importers in export markets often require the wine producer to have organic certification.

## Official labelling

The EU has laid down rules on organic farming and now also on vinification. Compliance with these rules is checked by a number of private firms approved in the country concerned. They carry out annual controls and every year issue a certificate permitting the producer to label his product as organically grown. The rules indicate which products may be used for spraying, which type of fertiliser and soil improvement is permissible and which additives are allowed in the must and the wine.

The EU adopted rules on vinification in February 2012, and as from the 2012 vintage organically certified growers must include the EU organic logo on their labels. The logo was 'democratically elected' on the web in 2010 and is now obligatory, just as it already was for other packaged organic agricultural produce (see p. 104).

Studies have shown that consumers attach importance to being able to tell the difference between organically grown and conventionally grown wines. In other words, good labelling matters.

In all EU countries, then, there are companies authorised by the government to control the organic growers, and it is these companies that issue organic certification. Growers in France most commonly belong to Ecocert, which controls about 80 per cent of the total number. Ecocert operates internationally and is represented in eighty different countries. Other approved control agencies in France are Qualité France, Agrocert, Certipaq and Certisud. Every organic winery is inspected at least once annually and in some cases up to four or five times. The inspections take place out in the field, where soil samples are taken for analysis warehouse stock is inspected, as are accounts, suppliers' receipts, and so on. Inspections can be both announced and unannounced. Certification costs about €300 a year.

*A certificate issued to Domaine Sainte Croix, Corbières, Languedoc, by one of the control agencies.*

EU grants are paid to all growers converting to organic growing. The grower undertakes to work organically for at least five years. The grant varies between €100 and €900 per hectare, depending on the crop cultivated. For wine-growers it is about €350 per hectare annually for five years (the three conversion years plus two more). The grant is meant to provide a measure of compensation for reduced yield during the year or years, because raising prices can be difficult.

## The whole EU procedure

Every year, a host of growers decide to convert their vineyards to organic growing. How do they go about it?

The registration process for a wine-grower is much the same everywhere in the EU. We take the example of a grower in France.

After careful consideration, a grower has decided to go organic. The first thing they do is notify the government body Agence Bio, which is tasked with developing and marketing organic agricultural produce and registering organic producers. They then decide which control agency (Ecocert, and so on) they want to join, and enrol accordingly. For three years their property will be 'under conversion'. In the second conversion year they can put *'en conversion bio'* (under conversion to organic growing) on their label.

The outcome of the conversion will depend on the starting situation, for instance, whether or not inputs of synthetic products have been gradually scaled down already. It is important to have the whole family and/or all employees on board. In order for everything to run smoothly, everyone must agree that they are now working differently. It is a good idea to meet and talk with others who have done the same thing or are also going through the conversion process. Some growers may need to go on a training course or perhaps call in a consultant or adviser. The soil may need analysing, so as to see what condition it is in and simplify the planning of work.

Briefly, then, the grower must:

🌿 notify the government body Agence Bio that they are converting to organic growing.

- enrol with a control agency (for instance, Ecocert).
- grow the vines without spraying them with synthetic chemical products and without using chemical weed killers.
- limit the use of additives in must and wine and refrain from certain techniques under the rules on vinification.
- not use genetically modified products.
- pass the control agency's annual inspections.
- put the control agency's code number on the label. In the case of Ecocert, for example, this will be: FR-BIO-01 (FR for France and 01 for Ecocert). This is obligatory. Every country and control agency has a unique code of its own. (Previously, just putting, say, 'Ecocert' was sufficient.)
- display the EU's organic logo, the little Euroleaf, on the label.
- notify Agence Bio annually that they are continuing with organic growing.

In the second conversion year they can put on their label that they are 'under conversion'.

It has also been permissible to declare on the label that the wine is made from organically grown grapes, for instance, *vin issu de raisins d'agriculture biologique*. As from the 2012 vintage, producers are instead to write that the wine is organic: *vin biologique*.

Most producers in France also choose to put *AB* (*agriculture biologique,* organic farming) on the label; this labelling comes under the French Ministry of Agriculture. The French are very familiar with it, and it helps many consumers to find organic products. Germany has a corresponding official certification mark called *Bio-Siegel*.

## The USA

The USA also has official rules on organic wines, and these also cover vinification, and so it has long been possible in the USA to speak in terms of organic wines. On the other hand, there are no more than about a dozen producers of such wine, because in the USA an organic wine may not contain any sulphur.

Under the National Organic Program (NOP) which was introduced in 2002 and comes under the US Department of Agriculture (USDA), there are two separate levels of organic wine, followed by a level for wine made from organically grown grapes.

- 100% Organic Wine: wine made from organically produced grapes only and with no sulphur added. In addition, the wine's natural sulphur content may not exceed 10 mg per litre. These wines may carry the USDA logo on their labels.

- Organic Wine: wine made from at least 95% organically produced grapes and not more than 5% non-organically produced, and with no sulphur added. The natural sulphur content may not exceed 10 mg per litre. These wines may also display the USDA logo on their labels.
- Wine made with organic grapes: wine made using at least 70% organic grapes. The wines may not display the USDA logo, but on the other hand sulphur may be added to them, and so most American organic growers end up in this category, despite using 100% organically produced grapes. Up to 100 mg sulphur may be added per litre. The same maximum limit applies to wines of every kind.

In the USA, then, an organic wine is made without any sulphur added. At present there are no more than a dozen or so producers who meet these exacting requirements and are allowed to put USDA Organic Wine on their labels. Many American wine producers are now bridling at this law. They grow organically and use 100 per cent organically produced grapes in their wines, but they add a little sulphur because they think it improves the wine. These growers do not like being lumped in a category which sanctions 30 per cent non-organic grapes. Many are converting to biodynamic growing instead and obtaining Demeter certification. That label, they feel, does them more justice. And Demeter allows sulphur.

## Switzerland

Switzerland too has official rules on organic growing. The certification mark is called Bio Federal and the rules are reminiscent of the EU's. In addition, most organic growers in Switzerland have the stricter private Bio Suisse certification label.

## Non-Certification

Some producers object on principle to having to pay for organic certification. 'It's the ones using toxins who ought to pay,' as many of them put it. The existing arrangement may seem unfair, but as long as the organic producers remain in a minority, it is hard to imagine any other solution.

Some growers attach little importance to being able to substantiate that they work organically or biodynamically. This is particularly the case with high-quality estates – such as Selosse in Champagne or Rayas in Châteauneuf-du-Pape – which get write-ups anyway and can sell their wines without the organic stamp. Domaine de la Romanée Conti in Burgundy, for example, had been producing organic wines for many years but only obtained certification in 2009. To their customers, presumably, it is of no consequence whether they work organically or not.

Some producers are put off by the administrative side of things, feeling they have enough red tape to deal with as it is. Organic certification, they argue, means more form-filling, still more controls, and so on.

Can one put trust in producers with no certification? Are principle, money and red tape the only reasons? Or do they want to have a small bolt hole in case they encounter problems in a difficult year? As consumers, we must each decide our own answers to those questions!

Nicolas Rossignol of Domaine Rossignol-Trapet in Gevrey-Chambertin, however, has another, interesting perspective on the issue of certification *vs.* non-certification: 'If no one obtains certification,' he says, 'nobody will know how many organic growers there are, and then we will have greater difficulty as a group in pressing demands on research, for example.'

*All official organic control agencies have a code which must be displayed on the label.*

*Experimental horse-drawn ploughing at Champagne Philipponnat's Clos des Goisses.*

*Fine-ground quartz for preparation 501.*

*Cow horn and various dried herbs.*

# CHAPTER 10

# Biodynamic Wine Production

*'I didn't believe in biodynamic production at first – it was tasting biodynamic wines that convinced me.'*
Frédéric Duseigneur

All biodynamic wine producers grow their grapes organically, but they also employ a number of other methods which are specific to biodynamic farming. So, in a manner of speaking, biodynamic growing is a variant of organic growing. Only a very small proportion of organic wine producers are biodynamic, but biodynamics has attracted widespread attention all the same, thanks to their number including some of the world's most outstanding and best-known wine producers. Altogether some two per cent of the world's organic wine producers are biodynamic.

Biodynamics is not always easy to understand, and biodynamic growers themselves would be the first to admit it. Rudolf Steiner, on whose thoughts biodynamics is based, was also the founder of anthroposophy. He was a student at the Vienna Technical University when in 1882, aged 21, he was tasked with editing the complete scientific writings of Goethe. Steiner felt an affinity to Goethe and his way of looking at nature, and Goethe is a constant thread running all the way through Steiner's life. The Goetheanum, for example, was the name given by Steiner to the anthroposophical movement's headquarters near Basle, designed by himself. The building, work on which began in 1923, is of cast concrete and ranks as an Expressionist masterpiece.

Without some knowledge of anthroposophy it is hard to read Steiner's books and assimilate what he wrote on the subject of biodynamic growing. Some growers converting their land, but by no means all of them, go on courses to learn more about Steiner and anthroposophy. Many concentrate their attention on the practical side of things and the resultant wine, paying little heed to the underlying philosophy.

Biodynamic wine producers try to make their vines strong and vigorous, so strong that they can resist attacks from diseases unaided. The vine, in their view, must defend itself. To help it, wine producers have nine 'preparations' which are central to all biodynamic thinking. They are used in the vineyard in homeopathic quantities, i.e. in very, very heavily diluted concentrations. Three of them are sprayed on the vineyard and six are blended into the compost, which is another important ingredient in biodynamic growing.

The biodynamic way of farming is described in Steiner's books. In 1924 he delivered a series of lectures to anthroposophical farmers on how the soil was to recover the health which, he maintained, it had lost due to the use of artificial fertiliser. The lectures were published in book form, enabling us to read Steiner's ideas on how the soil was to be reinvigorated and kept in shape. His books deal with agriculture generally and not specifically with wine production.

Many of the biodynamic methods can seem complicated when one reads about them. And peculiar. Steiner said that 'to our modern way of thinking this must sound utterly insane' when speaking about one of his preparations. But everyone puts their own spin on biodynamics. One should not believe that all wine producers follow Steiner's written precepts to the letter. Some of them skip one or two preparations, and it is quite common to buy at least some of them ready-made instead of making them oneself.

Biodynamics comes in for a lot of criticism because much of what biodynamic growers do cannot be explained scientifically. The growers are very much aware of this. Time and time again, one hears them saying the same: 'It's hard to believe in before you've seen it with your own eyes. It's only when you've seen the results that you realise it works.'

Some people have not been properly convinced until they have actually been able to compare exactly the same wine from organic vines and biodynamic ones. Anne-Claude Leflaive of Domaine Leflaive in Burgundy spent several years experimenting along these lines with a number of plots of vines on the estate. Not until she herself and others had opted for the biodynamic wine in several blind tastings did she decide to convert the estate entirely.

It was primarily the advent of artificial fertilisers in agriculture that Steiner objected to. The use of artificial fertiliser, he says, is not a good thing, because it is a 'dead' fertiliser with nitrogen, potassium and phosphorus. Everything added to the soil, he insisted, must come from living things – either the plant kingdom or the animal kingdom. He feared that if the use of artificial fertilisers was allowed to continue, farmland would be utterly impoverished (as indeed happened in certain wine-growing districts).

## The practical side

'Starting to farm biodynamically is a process,' according to Frédéric Duseigneur of Domaine Duseigneur in the southern Rhône Valley. 'It takes time, you proceed step by step, the feeling is important, and that is something not present in modern agriculture.' Most farmers converting to biodynamic growing are organic growers already, and as such are already familiar with this way of working. But biodynamics takes some getting used to all the same. 'At first you just follow the recipes,' Frédéric says, 'then you study the plants and begin to understand what you are doing, and then you can start experimenting on your own.'

## The preparations

The nine biodynamic preparations are central to biodynamics. It is above all these preparations that distinguish biodynamics from organic growing. They are referred to by their numbers, 500-508. This, it is believed, is a legacy from the Nazi period in Germany, when biodynamics was prohibited, so that people spoke secretly of numbers instead of mentioning the preparations by name. All preparations are composted by leaving them buried in the soil for six months or a year, two of them in cow horns and five of them in different animal parts (bladder, intestines, cranium). This is one of the hardest aspects of biodynamics to understand. Steiner himself says: 'Living things always have an inner and an outer aspect. The inside is encapsulated in a skin of some kind, the outside is outside that skin.' The preparations therefore have to be encapsulated in something while they are being composted.

The most important preparations are 500 and 501. These two work according to quite contrary principles. Both of them are buried in the soil in cow horn, 500 in winter, when the soil's forces are turned inwards, and 501 in spring, when life turns outwards again. Cow horn, according to Steiner, captures energy which can be sent back to the cow's digestive tract, and this energy can also be used after the horn has been removed from the cow.

Here we will describe the different preparations and their workings as posited in biodynamics.

## 500 – Horn Manure

Preparation 500 is made up in the autumn (that is, for those who want to make it themselves). However, not everyone can get hold of cow horn so easily, and some prefer buying the ready-made mixture. Once the cow horn has been obtained, it is filled with fresh manure from biodynamic and organic cows, preferably local ones. About one horn per hectare is required. The horns are buried in the soil, under several layers if necessary. All of them have to be surrounded by soil. Here they will remain during winter, for six months. When they are dug up again, what you have, according to Steiner, is 'an extremely concentrated, living manure'. The contents are diluted with water and mixed in a dynamiser before being sprayed over the vineyard. Steiner says that the same horn can be used three or four time. You spray with 500 at the onset of winter and perhaps a couple more times during the year, using about 200 g of the mixture per hectare (3oz per acre). The solution is sprayed on to the soil and ploughed in immediately afterwards. 500 makes the soil more porous, which helps the roots to grow longer and spread out. This way the vine is less liable to be damaged by drought and will have more mineral uptake.

Frédéric Duseigneur wondered if the use of cow horn was really

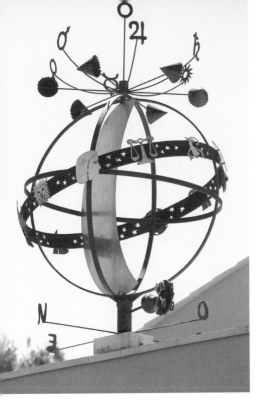

*A rooftop astrological globe at Domaine de Marcoux.*

*Preparation 500.*

necessary or if you could bury the cow dung in some other receptacle. He tried glass jars and yoghurt tubs, but it didn't work. 'When I dug jars and tubs up again,' he says, 'their contents were quite different from those of the cow horns.'

### 501 – Horn Silica

501 is made in spring, and the cow horns, now filled with fine-ground quartz, lie buried all summer. The quartz contains silica (a silicon compound), which conducts light and electricity, and the 501 helps photosynthesis to proceed as it should. The silica also reinforces the plant's immune defence. '501 is used in ridiculously small quantities,' says Frédéric Duseigneur, 'only 4 g per hectare (one twentieth of an ounce per acre), but it is a very powerful solution, replete with warmth that dries up the vineyard atmosphere and ensures good bloom and growth.' 501 is sprayed on the leaves in spring, preferably in the early morning, while the vine is blooming, and sometimes later in the growing season as well.

## 508 – Horsetail (Fr. prêle)

508 is a decoction of horsetail, a common plant which contains a lot of silica and thus helps the vine in the same way as the ground quartz. But horsetail also contains sulphur, and so 508 serves as a fungicide too, inhibiting the development of fungal diseases on the vine. Horsetail, says Frédéric Duseigneur, is useful above all in damp, northern districts, because it dries out the ground. There is a reason, he maintains, for its often growing in damp places: the roots pump up surplus water.

The decoction is dynamised (see below) and then sprayed on the vines, drying them out. Frédéric Duseigneur likens it to a vacuum cleaner collecting the moisture. Emmanuel Cazes at Domaine Cazes in Roussillon also believes that horsetail makes the grapes thicker-skinned, which is all to the good in a damp climate, because then the grapes will be less prone to rot.

## The dynamiser and dynamisation

500, 501 and 508 all have to be dynamised before they are sprayed on the vineyard.

The dynamiser is a round container filled with water to which the preparation is then added. The entire contents are stirred, first in one direction, then quickly changing to the other. This agitation creates a 'vortex', and the change of direction creates a 'chaos'. Both of these, according to biodynamics, are important stages. Water is considered to have a 'memory' and so it will memorise the power of the biodynamic preparation and then pass it on to the vineyard. Stirring takes between 20 minutes and an hour altogether, depending on the mixture involved. It can be done manually or mechanically.

There is also another kind of dynamiser instead of the round one in which one stirs the contents. This is a structure resembling a stepped cascade sometimes called a 'flowform'. The water flows down continuously through a sequence of 'bowls' which are shaped in such a way that both 'vortex' and 'chaos' are created as the water runs down through them.

The power needs to be swiftly passed on to the vineyard,

*Frédéric Duseigneur sniffing his compost.*

*Compost, with small fungi, at Villa Bellini, Valpolicella.*

*Willow leaves.*

*A traditional dynamiser at Domaine Humbrecht.*

*A flowform dynamiser at Domaine Turner Pageot.*

within two or three hours of dynamising, and so the spraying has to be done directly afterwards. Accordingly, the preparations are saved till spraying time. Only then are they mixed with water and dynamised.

## The compost

The soil must be vitalised – this is a fundamental tenet of biodynamics. The soil must be given strength and nourishment the natural way, partly through biodynamic compost. Animal dung is used, preferably from cows but also from other animals, not uncommonly from sheep. Preferably the dung should come from one's own livestock. Theoretically, a biodynamic farm should be self-sufficient and should not need to bring in anything from outside. This, of course, is impossible for most wine producers, especially in Europe. South American producers have more land. The big Emiliana estate, Chile's only Demeter winery, has its own herd of cows, 'so that we can be sure that the grass they graze on is organic.'

Ground-down branches, dead plants, grape skins, pips and so on, are mixed together in the compost. Ingredients from the plant and animal kingdoms must be judiciously balanced. The compost is carefully watered and stirred. 'It's left to ferment for five months,' says Frédéric Duseigneur, 'and in autumn it is spread in the vineyard. By then the quality is so good that only small amounts are needed.' One ought preferably to make one's own compost, but failing that, one can buy compost from a biodynamic supplier.

The growers themselves say that biodynamic compost has superior drainage properties, is less compact and contains more worms and nourishing fungal mycelium. Some research findings point in the same direction, though as yet only tentative conclusions are possible.

Georg Meissner is researching this very topic at Geisenheim University, Germany, and when pressed on the subject of scientific proof he cautiously replies: 'These are difficult questions. For five years now we have been comparing the three systems (conventional, organic, biodynamic) in a vineyard planted with Riesling, but we have not yet published any findings. Research has to go on for a long time – people often publish prematurely. But things are looking good for biodynamics, certain tendencies are visible. The bunches of grapes are less compact.'


## Compost turns biodynamic

Steiner said that 'the fertiliser of the future should be made, not with chemicals but with yarrow, chamomile, oak bark and dandelion. That kind of fertiliser will contain everything the plant needs.' And, he went on, if valerian is then added, this will improve matters still further. These preparations are 502-507, and the first five are prepared by leaving them buried in the soil. Ideally all of them should be added to the compost, but the odd one can be dispensed with.

## Preparations 502–507

### 502: Yarrow (Fr. Achilée)

Yarrow contains potassium and selenium, even in soils otherwise containing very little of these minerals. Steiner found it a fantastic plant, bringing health to the vineyard by simply growing there.

502 is made by picking the flowers and putting them out to dry, after which they are put inside a deer's bladder. This is hung up in a sunny place for the summer. When winter comes it is buried in the soil and left there for the entire season. Next the content is removed and added to the compost. The energy from this preparation, Steiner tells us, is so great that, no matter how little one adds, it will affect the compost. The deer, Steiner maintains, is quite a different sort of animal from the cow. Its antlers point outwards, towards the earth's surroundings, the cosmos, whereas the cow's horns point inwards. The bladder is a sensitive organ, 'almost a copy of the cosmos,' according to Steiner, and biodynamics has it that by putting the yarrow in a deer's bladder we make the vines and their fruit more sensitive and receptive to cosmic influence.

### 503: Chamomile (Fr. Camomille)

Chamomile vitalises the vine. It contains calcium and nitrogen. It stabilises the nitrogen content of the compost. The flowers are dried and made into little sausages, using cattle intestines. The sausages are hung up all summer and dug down in winter.

### 504: Stinging nettles (Fr. Ortie)

Steiner writes that nettles 'stabilise the nitrogen content of the soil'. They also help the soil to convert organic substances into minerals. This is the easiest preparation to make. Freshly picked stinging nettles are buried in the soil and left there for a year. Failing fresh nettles, you buy dried ones. Emmanuel Cazes of Domaine Cazes, Roussillon, points to the high concentration of iron in nettles. 'Nettles,' he says, 'also improve the flow of sap in the vine.'

## 505: Oak bark (Fr. Écorce de chêne)

Steiner writes that oak bark puts calcium into the soil and that calcium 'helps to drive away plant diseases'. This is a complicated preparation to make. Grated oak bark is put into the cranium of a sheep, pig or horse and buried in autumn and winter. It is then dug up again, dried and added to the compost. Oak bark adds calcium to the compost and improves the vines' immune defence.

## 506: Dandelion (Fr. Pissenlit)

Steiner regarded the humble dandelion as a messenger from heaven. The flowers are picked and left to dry a little before they are sewn into the mesentery of a cow. The package is buried in the soil in winter, and when lifted it is, according to Steiner, 'replete with cosmic influence and will help the plant to absorb exactly the amount of silicic acid it needs.'

## 507: Valerian (Fr. Valériane)

The flowers are pressed and put to soak in plenty of water. Valerian stimulates phosphorus activity.

'Together,' Steiner wrote, 'these biodynamic preparations revivify the soil, fill it with nourishment and assist with organic degradation.'

In France, the biodynamic preparations are not classed as PNPPs (see Chapter 8). At present, according to Jean-Marie Defrance of Demeter France, they do not have any official status. Discussions on how to classify them have been ongoing ever since the outbreak of mad cow disease, he says, adding that the preparations will probably end up being classed as 'composting aids'.

Many biodynamic growers make other plant-based preparations of their own, in addition to 502–507. Continuous experimentation and the devising of new recipes capable of assisting the soil and the crop grown are part and parcel of the biodynamic mindset.

## Cosmic rhythm, the moon, and four kinds of day

Most biodynamic growers pay some deference to what Steiner called 'cosmic rhythm' when carrying out their various tasks in the vineyard. Steiner himself attached great importance to this, maintaining that modern science was too narrow-minded and failed to take into account the 'cosmic forces' imparting energy to plants, which then pass it on to us humans.

To help them, growers have a planting calendar which is drawn up every year, the best known one being *The Maria Thun Biodynamic Calendar*, though other people also compile planting calendars. These calendars show the positions of the stars and planets in relation to the fixed constellations and indicate the best days for a particular task in the vineyard. Performance of that task may be appropriate or inappropriate, depending on whether it is a root, leaf, flower or fruit day. But Emmanuel Cazes of Domaine Cazes, Roussillon, finds the planting calendar more effective where non-perennials are concerned. Different growers heed the calendar to differing extents.

Maria Thun, who died in 2012, worked all her life within a small area of Steiner's indications. To study the effect of the moon and planets on plants, she planted radishes on different days to see if it made any difference, and there did prove to be a difference, in both taste and appearance. Some radishes turned out bitter, others spicy, some grew well, others not so well. These experiments prompted her to start compiling the biodynamic planting calendar.

By sowing, harvesting, ploughing, planting, and so on, on the 'right' days, one can stimulate the growth potential and health of the vines. During an ascending moon, for example, there is more moisture in the soil and one must beware of the greater risk of fungal attack. But this is also a good time for spreading compost. When the moon is in the descendant, the roots develop more easily, and so this is an opportune moment for planting.

Biodynamic wine producers, however, are not the only ones to take account of the moon. It is not uncommon today for 'ordinary' organic or conventional growers to say that they take the phases of the moon into account as much as possible, for example, when racking or bottling their wine or planting new vines. A conventional producer can, for

*Field horsetail at Champagne Tarlant.*

*Yarrow.*

*Dried orange peel can be used in biodynamic preparations.*

example, say that 'in the right moon phase the lees are far more stable on the bottom of the tank, and so this is the right time for racking.'

## The consultants

Many people have continued disseminating and developing Rudolf Steiner's teachings. Biodynamics was by no means a cut-and-dried doctrine bequeathed by Steiner to posterity when he died in 1925, a year after giving his famous lectures. On the contrary, he left many question marks after him. As we have seen above, **Maria Thun,** of course, has been a leading figure. She helped growers with her sowing and planting calendar, and she also developed preparations of her own, for instance, the widely used Maria Thun barrel compost, consisting of cow manure and some of the plants from the preparations. It is spread in the vineyards during winter, in some cases up to three times.

The best-known French biodynamic wine consultant, unquestionably, was **François Bouchet** (died 2005). He assisted many famous estates: Leroy and Leflaive in Burgundy, Chapoutier and Montirius in the Rhône Valley and Pierre Frick in Alsace, to mention just a few. His own estate in the Loire Valley, the five-hectare Domaine de Château Gaillard, now run by his son Matthieu, was already converted by François in the 1960s. François's daughter Véronique runs a biodynamic estate in Bordeaux, Château Falfas. François Bouchet's book *L'Agriculture Biodynamique* is a widely read introduction to biodynamic farming.

**Alex Podolinsky** is a biodynamic grower in Australia. He has been, and still is, active as a wine production adviser, not only in Australia but in Italy, France and Switzerland too. He is one of those who have created preparations of their own. With his simplified 500P preparation, for example, he wants to make it easier for big estates to go biodynamic. He registered the Demeter trademark in Australia, thereby preventing Demeter International from becoming established there under that name. Alex Podolinsky has devoted a long life to biodynamics. He started in the 1950s (he was born in 1925) and is still going strong.

His 'disciples' include the Frenchman **Pierre Masson** who runs

*Frédéric Duseigneur harvesting oak bark.*    *Dried dandelions at Champagne Tarlant.*

*Dried chamomile.*

a firm of consultants (Biodynamie Services) in Burgundy. Pierre Masson writes books, publishes a biodynamic calendar and sells the biodynamic preparations, dried plants for decoctions and variously sized copper dynamisers.

Other well-known consultants include Jacques Mell, who has many clients in Champagne, Alan York, who has worked for a long time with Bonterra in California, and Peter Proctor in New Zealand.

## Control and rules

Many biodynamic wine producers are certified by Demeter, the leading organisation for biodynamic farming, headquartered in Germany. Many countries have local Demeter associations, and these are what growers belong to. One can only belong to Demeter International if one lives in a country without any local Demeter organisation.

Some French growers opt instead for membership of the Biodyvin association, the full name of which is Syndicat International des Vignerons en Culture Bio-Dynamique. Founded in 1990, it now has roughly seventy-five members.

Both Demeter and Biodyvin have rules on growing and vinification. They specify what may be used in the vineyard. Demeter France and Demeter Austria, among others, have limited the use of copper to 3 kg per hectare (2.5lb per acre) annually.

Demeter and Biodyvin growers are expected to aim for minimum use of additives in their vinification. Some are tolerated, but the aim must be to keep on reducing the inputs. Fermentation must be done using natural yeast. Organically grown grapes are considered to have enough yeast on their skins already and for the most part to present no problems. Any additives, such as sugar for chaptalisation or whites of eggs for fining, must be of organic origin.

Acidification and the addition of enzymes or tannins are prohibited for wines of every kind. Bentonite is permitted for fining white wines and the whites of Demeter-certified or organic eggs for fining red ones. Cellulose and tangential filtration is permitted, but centrifuging is banned. Reverse osmosis and gum Arabic are forbidden for all wines.

*Storage of the 500P preparation (a variant of 500).*

*The number to call if you want to plough your vineyard with animals instead of a tractor. Champagne Philipponnat.*

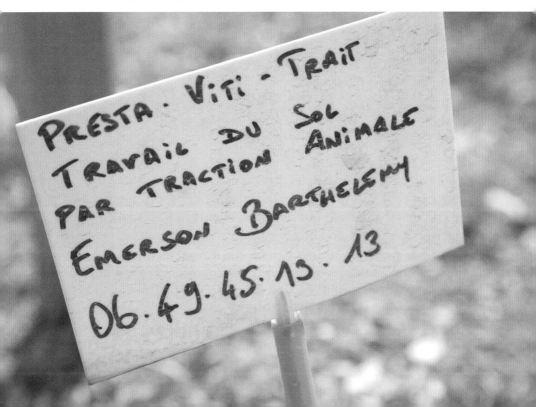

*Nicolas Joly, Domaine de la Coulée de Serrant.*

*Noël Pinguet, Domaine Huet.*

Only Demeter or organic spirit may be used for the fortified *Vin Doux Naturel (VDN)* wines from the south of France. Fermentation tank temperature control and cold stabilisation are allowed. Sulphur levels are to be kept down.

Demeter and Biodyvin both highlight prohibition of the additives PVPP (Polyvinyl-polypyrrolidone), gelatine, blood, fish glue, sorbic acid and ascorbic acid (further to this, see Chapter 13 on additives).

## Wine-growers on the practical results of biodynamics

Biodynamic methods arouse widespread scepticism, indeed sometimes almost hatred. It is important to understand that not all 'rules' are slavishly complied with. As we have now seen, many of Steiner's followers have added preparations of their own, and besides, work has to be adapted to climate and other extraneous influences. In South America and Australia, for example, some of the preparations

*Olivier Humbrecht, Zind Humbrecht.*  *Frédéric Duseigneur preparing a herbal infusion.*

proved hard to make because the plants did not grow there. So the preparations have to be adapted to local conditions.

Either way, there is no questioning the enthusiasm of many converts to biodynamic growing, as the following comments show.

Emmanuel Cazes, Domaine Cazes, Roussillon: 'The soil became darker, acquired more humus content, and the acidity of the wine increased, the wines, quite simply, took on a fresher taste. With biodynamics the vine can cure itself, but it prefers not getting ill in the first place.'

Jean Pierre Frick, Domaine Pierre Frick, Alsace, finds that biodynamic farming transforms the vineyard. 'The vines are more resistant and less vulnerable to grey rot and grape worm attacks. It is easier getting the grapes to ripen well and getting more minerality in the wines. In addition, the wine becomes more concentrated.'

Michel Comps, Château Les Mangons: 'I met people who were farming biodynamically, and I got interested. I didn't want to go on as I was, crop spraying. I felt that more and more interventions were becoming necessary.' He finds biodynamic farming different and more

positive. He converted the whole of his acreage, 18 hectares, in one sweep, and the process was not altogether painless. The yield declined at first, but it steadied, and he feels that the wines now have a more pronounced minerality and a better *terroir* style.

Christine and Eric Saurel of Domaine Montirius in Vacqueyras, in the southern Rhône Valley, noticed results from the very outset, and can still be surprised to see how well their vines are prospering. Take the summer of 2011, for example. There was extremely little rainfall and it was very hot, but their vines did not suffer from any hydric stress at all. They were healthy and green, a sign, Christine and Eric maintain, that they were getting enough water. 'We believe it is thanks to our sixteen years of biodynamic farming that the vine roots reach so far down, far enough to get all the water they need.' Christine hastens to add that flexibility is essential. For example: 'You can't spray if the Mistral is blowing, even if it's a perfect fruit day.'

*Herbs – thyme and rosemary – drying at Domaine Duseigneur.*

# CHAPTER 11

# Private Labelling and Control

Wine-growers can have many reasons for deciding to join a private organic association. Perhaps they do not think the official rules are strict enough, and indeed, many countries do not have any official rules. Then again, very often the private association can help with marketing and advisory services. Another important consideration for growers in the EU countries has been that the private organisations have rules on vinification, which the EU did not have until August 2012. Many growers want to be able to show that they also comply with organic principles inside the wine cellar. Out in the field, the official rules (EU rules in the case of an EU country) are mandatory, but private certification labels can have stricter rules if they want to, for instance, on the amount of copper permissible.

What is permissible or not permissible varies to some extent from one association to another. Sometimes there is consensus on what is prohibited, but not always. The basic principle is to do as little as possible in the cellar, keeping intervention to a minimum. If the vineyard has been properly tended, one will have had fine grapes to pick, and should therefore have no difficulty in producing a wine without additives. But nearly everyone uses some kind of 'additive', the only question being which ones and in what quantities. We have often been told in recent years that fine grapes are the most important quality factor. Work in the cellar must be 'non-invasive', with the wine 'making itself', and so on. This is not wholly true. However natural one may be, it is man who makes the wine, not Nature. And most producers do not consider the use of modern inventions like temperature control, *remontage* (pumping over) and *pigeage* (punching down the cap), and so on, at all contrary to organic thinking.

*Ripe grapes in a Campania vineyard.*

A serious organic grower naturally tries to maintain the same philosophy indoors as when working outside in the vineyard. Using additives as sparingly as possible is second nature to them. How they then choose to make their wine is very much a question of what type of wine they are out to make and for what market.

## Private organic labels and organisations (a selection)

There are roughly a hundred different organic certification marks in the world today, but we can console ourselves with the fact that not all of them are concerned with wine. They have all developed since the late 1940s, when they were quite simple in structure. The members checked one another, in so far as there were any controls at all. Nowadays controls for practically all certification labels are carried out by a third party.

## International

### The International Federation of Organic Agriculture Movements (IFOAM)

The International Federation of Organic Agriculture Movements (IFOAM) was founded in 1972 as an international umbrella organisation for organic farming, and acts as a kind of co-ordinator or helpmeet. It has a list of products which it regards as permissible in organic farming, and the list can be consulted by countries or private organisations. IFOAM tries to simplify the co-ordination of organic rules worldwide. The IFOAM certification mark on the wine bottle shows that the vineyard conforms to IFOAM's international guidelines.

## France

### Nature et Progrés

Nature et Progrès was formed in 1964 in protest against industrialised farming and today is one of the oldest organic organisations in Europe. It unites producers and consumers wishing to promote an agriculture which respects 'living things' and who wish to develop organic farming. Strict rules apply to both growing and vinification.

### Fédération Nationale Interprofessionnelle des Vins de l'Agriculture Biologique (FNIVAB)

FNIVAB was founded in 1998 and since 2003 has also had rules on vinification. At present the organisation has some sixty members all over France, and they are entitled to include the FNIVAB logo on their labels.

## Vin Bio d'Alsace

In response to the EU's failure in 2010 to agree on rules of vinification, the Alsace growers took matters into their own hands and drew up rules of their own. Vin Bio d'Alsace was presented in Kientzheim in May 2011 and today has forty or so members. These pledge themselves to observe the same philosophy inside the cellar as out in the fields. Sulphur inputs were limited to 120 mg/l for dry wines, that is, wines containing less than 5 g sugar per litre, and to 100 mg/l for red wines. These limits were, respectively, 80 mg/l and 50 mg/l below the rules applying within the EU at the time.

It is uncertain at present whether Vin Bio d'Alsace will remain in existence now that the EU has organic wine.

# Austria

### Bio-Austria

This is the biggest private certification label, representing four hundred growers. It is important to note than only 3 kg copper sulphate may be used per hectare annually, that is, only half the amount sanctioned by the EU rules.

# Switzerland

### Bio Suisse

Switzerland's biggest and best-known organic certification label. Known as *Le Bourgeon* (the bud), it does indeed resemble a tiny flower bud. Bio Suisse has strict rules on vinification.

# Italy

## *Associazione Italiana per l'Agricoltura Biologica (AIAB)*
AIAB was founded in 1982 as 'Commissione nazionale cos'è biologico' and changed its name to AIAB in 1988. AIAB issued Italy's first organic farming standard in 1992. The association also has rules on wine production.

## *Istituto per la Certificazione Etica Ambientale (ICEA)*
This organisation certifies organic farmers and wine producers in and outside Italy.

# Germany

## *Ecovin*
Founded in 1985, Ecovin is Germany's biggest private organisation for organic wines, representing just under half of all organic wine producers in the country.

# California

## *The California Certified Organic Farmers (CCOF)*
Has been certifying organic food products, including between twenty-five and thirty wineries, since 1973. Through the CCOF one can also obtain USDA certification and then use both certification marks.

106 090067

LSSR 091
STANDARD DRINKS

50MLS

CERTIFIED
bio·org
ORGANIC
1517

14

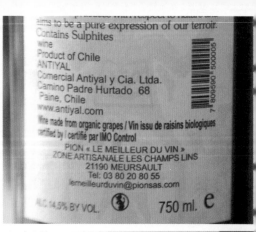

aims to be a pure expression of our terroir.
Contains Sulphites
wine
Product of Chile
ANTIYAL
Comercial Antiyal y Cia. Ltda.
Camino Padre Hurtado 68
Paine, Chile
www.antiyal.com
Wine made from organic grapes / Vin issu de raisins biologiques
certified by / certifié par IMO Control
PION « LE MEILLEUR DU VIN »
ZONE ARTISANALE LES CHAMPS LINS
21190 MEURSAULT
Tel: 03 80 20 80 55
lemeilleurduvin@pionsas.com

ALC 14,5% BY VOL.          750 ml. ℮

7 809590 500005

Weingut Sander ∗ D-67582 Mettenheim
Erzeugerabfüllung ∗ A.P.Nr. 4 300 133 03 11
DE-022-Öko Kontrollstelle ∗ Enthält Sulfite
Product of Germany ∗ Deutscher Qualitätswein

Wein aus ökologisch
erzeugten Trauben

demeter

% vol
5 l

RHÔNE VILLAGES LAUDUN

010

BIO-DYNAMIQUE

BIODYVIN

incomparable terroir des éboulis du...

1% Vol     Mis en bouteille à la Propriété
par EARL DOMAINE GIACHINO
F38530 Chapareillan - Tél/Fax : 04.76...
www.domaine-giachino.fr
NATURE &
PROGRES
contains des sulfites

Vin issu de raisin cultivé en Agriculture...
Certifié par ecocert...

DEMETER
VIN ISSU DE RAISINS DE...
VIN ISSU DE RAISINS...

ouuce of Germany

750 ml
alc.12,5% vol

146     VDP.Prädikatsweingut     Naturland

Certified
Organic
Bio-org
BDOCA0899

# Australia

### The Biological Farmers of Australia

This is Australia's foremost association of organic producers. Certification (both organic and biodynamic) is handled through Australian Certified Organic (ACO). The rules, on the whole, are modelled on Europe's. Five per cent of the organic property has to be set aside for woodland or meadowland.

# South Africa

South Africa's organic wineries are often certified through foreign control agencies such as the Norwegian company Debio. These control firms are accredited to various European or American certification labels (EU, Bio Suisse, NOP).

# Chile

### Certificadora Chile Orgánico (CCO)

Founded in 1998 to promote awareness and development of organic farming. A certification committee is responsible for inspections and certification.

Just as in South Africa, Chilean wine-growers can also be certified by international control companies.

# Argentina

Argentina has several different organisations handling the labelling of organic foodstuffs:

- Argentina Organic Regulation
- Argencert Organic Standard
- Letis IFOAM Standards
- Oia Organic Standard

# New Zealand

### The New Zealand Biological Producers and Consumers Council (Bio-Gro)

The biggest organisation in the country for certification. Bio-Gro can also award certification to the EU, NOP (USA) and IFOAM standards.

## Biodynamic certification labels

All biodynamic labelling presupposes that the producer has organic certification.

### Demeter International

Demeter International is head-quartered in Germany. There are national Demeter associations in several countries (Demeter France, and so on).

Demeter International has only had rules on vinification since 2008, and they are continuously changing.

Anthroposophy does not encourage consumption of alcohol, and so certain other Demeter farmers were sceptical at first about making Demeter labelling available to wine producers. Due to this attitude, wine producers have not always felt comfortable with Demeter, which perhaps was one reason for the foundation of France's Biodyvin.

Demeter requires that all the biodynamic preparations be used (500 and 501 at least once a year), and all organic fertiliser has to be treated with all the compost preparations.

Demeter maintains that 'a wholly satisfactory result can be expected only if all preparations are used all the year round, in the right way and after proper dynamising.'

## Demeter France

Demeter France has vinification rules of its own which are somewhat stricter than Demeter International's.

## Demeter Association Inc. USA

'Biodynamic' (with a capital B) is a registered trademark in the USA and the property of Demeter Association Inc. One rule specific to Demeter USA is that not less than 10 per cent of the property's total acreage must be maintained as a biodiversity set-aside, for instance, as woodland, copses, meadows, wetlands, hedges or suchlike. There are two different certification labels to choose from:

- *Biodynamic wine / Demeter wine / Demeter certified wine:*
  Grape production conforms to the biodynamic rules
  for agriculture, and very few additives are permitted
  in the cellar (sulphur not more than 100 mg/l for all
  types of wine, no additional yeast, no chaptalisation, no
  acid adjustment, though on the other hand oak chips
  are allowed!). This status, the organisation says, may
  not always be attainable, dependent as it is on what the
  weather has been like during the year.

- *'Wines made from biodynamic grapes' or 'Wine made from Demeter certified grapes':*
  Subject to slightly less stringent rules of vinification.
  Additional yeast is allowed, so long as the yeast is not
  genetically modified (written certification is required
  from the supplier). Acid adjustment can be tolerated but
  must be documented (adjustment with ascorbic acid,
  citric acid or tartaric acid). If chaptalisation is done,
  biodynamic sugar or biodynamic concentrated grape
  must has to be used. The addition of tannins from the
  seeds and skins of biodynamic grapes is also permitted.

## Biodyvin France

Syndicat International des Vignerons en Culture Bio-Dynamique (Biodyvin) is a French association of biodynamic wine producers.

It was founded in 1998 and members must set themselves the aim of avoiding all additives to the wine which can change the original balance of the grape and remove the character of the vintage. It is important in the cellar to preserve the 'energy' which the grapes have derived from biodynamic cultivation. The organisation concedes, however, that it is not always easy to dispense with everything and that use of certain products may sometimes be necessary. This is tolerated if it can be justified and provided that the grower simultaneously experiments with techniques making it easier to avoid or reduce inputs of these products. Seventy-three wine producers with a combined acreage of nearly 1,700 hectares now have BIODYVIN certification.

In its rules, BIODYVIN specifies, as a minimum, that wine-growers must use:

- Preparation 501 between once and three times after harvesting.
- Preparation 500 between once and three times before budding.
- Preparation 501 once to three times after budding.
- Stinging nettles and horsetail as fungicides.
- Stinging nettles as insecticides, and also that:
- Copper, sulphur, bacillus thuringiensis and rotenone may only be used when absolutely essential.

Before a producer can be admitted to membership of Biodyvin, their wines have to be tasted and approved. Biodyvin is said to turn down many applications. Demeter does not have that procedure.

## Biodynamic Research Institute, Australia

This institute, founded in 1957 by Alex Podolinsky, is Australia's Demeter certification agency. Podolinsky registered Demeter as a trademark in Australia, which bars Demeter International from using the word Demeter in that country.

## La Renaissance des Appellations

La Renaissance des Appellations is not a labelling but a loose-knit association of some two hundred wine producers the world over, most of them biodynamic, started by Nicolas Joly. The association organises joint events and wine tastings.

Members of La Renaissance des Appellations must endeavour to make wines with a sense of place. That, Nicolas Joly insists, is the whole point of *appellations*. But, he continues, this sense of place or *terroir* has disappeared as a result of all the chemical spraying. His fundamental aim in starting the association was that one should always be seeking improvement, that is, endeavour to minimise the use of additives and interference with the wine.

# CHAPTER 12

# The Work Inside the Cellar

'Wine made from organically grown grapes' sounds wordy and stilted in the everyday context, but until recently it has been the correct name because there were not any official rules governing work inside the wine cellar. As we have already seen, that changed in Europe with the 2012 vintage. Of course, there have been rules concerning permissible additives and production processes in wine-making, but those rules applied to all growers, both organic and conventional.

EU rules governing cellar work on organic wines should have been introduced already in 2010. In 2006 the EU started a research project, ORWINE, aimed at working out standards for 'organic wine'. ORWINE worked till 2009 on drafting a proposal based on research and market surveys, and that proposal formed the basis of discussions between the different countries and the European Commission in 2010, but there were many differences needing to be resolved. The main stumbling block was inability to agree on the extent to which permissible sulphur levels were to be lowered. Work and discussions between the different parties and countries continued, and we now have an agreement, effective from August 2012.

## Why have rules in the cellar?

Must we also regulate operations in the cellar? Yes, there is a desire for this on the part of consumers, wine merchants and growers for the whole process to be controllable and for the possibility of calling one's wines 'organic', as in the USA. Among other things, it is widely believed that this would boost exports. But, as already commented, organic producers are divided on these issues, even within the different EU countries. Some feel that one should add as little as possible inside the cellar, basically nothing at all, and, moreover, that the majority of mechanical processes should be banned. According to ORWINE's studies, a surprising 27 per cent of the producers approached in Spain, Portugal and France felt that no additives should be allowed, not even sulphur. One finds it a little hard to believe that they were in earnest, given that very few growers dispense entirely with sulphur, even in these countries.

Consider, for example, the situation in the USA, where 'organic wine' may not contain any sulphur additive. Extremely few producers have taken that path. The majority do not want to manage without sulphur, nor do they wish to dispense with all other additives. Not surprisingly, it is growers in northern Europe who see the greatest difficulty in going without. The colder and rainier one's region, the harder it becomes to make wine without, for example, sulphur. In Germany, many organic growers wanted to retain the permissible sulphur content as for conventional producers.

But isn't sulphur a natural product which ought to be readily permissible for organic wines? Many people think so. Others do not attach very great importance to sulphur being a natural product, they are more concerned about it being an extraneous additive. The debate is further complicated by sulphur naturally forming in the wine during vinification, so that even if none is added, it can still be naturally present in the wine. Another complicating factor is that, even though sulphur is a natural substance, most if not all the sulphur used in wine production is derived from oil refinery waste products.

## Why add sulphur anyway?

The use of sulphur in wine production is practically unavoidable, and even the majority of organic growers concede as much. Sulphur is added above all to stabilise the wine and keep it from oxidising. Sulphur is an antioxidant, and as such keeps the wine from oxidising prematurely. White wines are more sensitive to oxidation than red ones, because they can change colour and darken, rather like a quartered

*Dropping a sulphur pastille into the full barrel.*

*Sulphur solution.*          *Sulphur pastilles.*

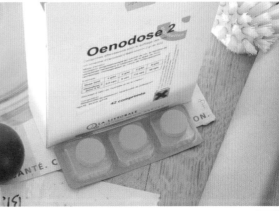

apple. Often, therefore, more sulphur is added to white wines than to red ones. With red wines the opposite applies – you have to be careful, because sulphur can make the colour turn lighter, which is never desirable where red wines are concerned. Red wines need less sulphur all the way, because they have tannins to protect them from oxidation.

Sulphur also acts as a microbiological stabiliser. A little sulphur added at the end of the production process will kill off any remaining yeast and thus prevent secondary fermentation or tainting in the bottle.

Only a few organic and biodynamic growers abstain completely from using sulphur. Some of them have successful results, others less so. Producing wine without any sulphur whatsoever is not an option for the majority of organic producers, but many of them aim to maintain lower levels than conventional growers. Standpoints on this issue vary according to the type of wine produced and the prevailing climate, and also, not least, on the quality of the vintage.

Consumers are often opposed to the use of sulphur and do not always realise that it is an additive which is (almost) ever-present. And not only in wine but in many other foodstuffs (for instance, dried fruit). Many consumers look on sulphur as something hazardous which causes headache and makes the wine foul-smelling and evil-tasting. Many think sulphur should be banned completely where organic wines are concerned. Statements of this kind are often rooted in ignorance and perhaps prejudice, too.

One can be allergic to sulphur – a small percentage of the population are and have to be careful. But a small, judiciously proportioned amount in wines will neither cause headache nor make the wine foul-smelling or evil-tasting. The prejudice against sulphur may possibly be due to the sulphur content of white wines especially having once been far higher. Most producers – both organic and conventional – are now using less. The EU rules can be seen as quite generous, as most wine producers agree (though with certain exceptions in Germany and Austria). One interesting point is that BiB (bag-in-box) wines need more sulphur additive than bottled ones, due to the wine keeping better in a bottle than it will in a box.

## Reduce by how much?

Prior to the 2010 vote, the EU's ORWINE project team had recommended cutting the maximum concentrations by 50 mg/litre from what was then permitted. This reduction was rejected by German and Austrian wine producers.

German producers wanted no reduction at all. That way, there would have been no difference in sulphur content between a conventional wine and an organic one. The reason for the German standpoint is that in a cool climate vineyards are more severely beset with fungal diseases. If infected grapes get into the fermentation tank, sulphur will prevent the infection reaching the must. But many producers in warmer countries were also diffident about reducing concentrations too much, because this can mean problems in rainy years.

Apparently, the EU could not contemplate having different rules for different countries. The same maximum limits would have to apply to all countries and districts. Perhaps it might be reasonable to consider having different rules for different districts, rather like today's (conventional) rules on chaptalisation (sugar additive), which sanction different amounts for different regions.

On the other hand, consumers must be able to identify an organic wine easily and rely on conformity to certain rules which are the same for everyone. And there are no inter-regional differences concerning, for instance, the amount of copper an organic grower may use in the vineyard, even though a northern climate may be more prone to downy mildew.

ORWINE's studies showed the majority of organic producers in Spain, Italy and France to be of the opinion that good wines could be produced with sulphur content well below the maximum limits applying to conventional wines. A 20 or 30 per cent reduction was judged appropriate. The German producers were of quite a different opinion. 70 per cent of them wanted to keep the conventional limit values, especially for dry white wines.

The new rules eventually agreed on meant a 30 mg per litre reduction compared to what is now permissible for conventional wines. At the same time, an additional category has been introduced for wines containing less than 2 g residual sugar per litre, a category which does not exist for conventional wines.

## Red wines

| Sulphur addition | EU rules for conventional wines | ORWINE's proposed new rules, 2010 | Proposal adopted by EU, 2012 |
|---|---|---|---|
| Red wines <2 gram residual sugar/l | | | 100 mg/l |
| Red wines <5 gram residual sugar/l | 150 mg/l | 100 mg/l | 120 mg/l |
| Red wines >5 gram residual sugar/l | 200 mg/l | 150 mg/l | 170 mg/l |

## White wines / rosé wines

| Sulphur addition | EU rules for conventional wines | ORWINE's proposed new rules, 2010 | Proposal adopted by EU, 2012 |
|---|---|---|---|
| White and rosé <2 gram residual sugar/l | | | 150 mg/l |
| White and rosé <5 gram residual sugar/l | 200 mg/l | 160 mg/l | 170 mg/l |
| White and rosé >5 gram residual sugar/l | 250 mg/l | 210 mg/l | 220 mg/l |

## Sweet wines

| Sulphur addition | EU rules for conventional wines | ORWINE's proposed new rules, 2010 | Proposal adopted by EU, 2012 |
|---|---|---|---|
| Sweet wines, no botrytis* | 300 mg/l | 250mg/l | 270 mg/l |
| Sweet wines (botrytis-affected)* | 400 mg/l | 350 mg/l | 370 mg/l |

(*The appellations and types of wine affected are specified in the EU rules.)

## The disadvantage of reducing sulphur content

But the sulphur-or-no-sulphur issue is a tricky one. Countries opposing the reduction claimed that it would change the character of the wine and might necessitate different working methods, for instance, using yeast additive instead of wild yeast, to offset the lower sulphur content. In addition, the risk of Brettanomyces ('Brett') attack, which at worst can makes wine fiercely malodorous, is increased by the abolition or drastic reduction of sulphur inputs.

The ORWINE report also showed several wine producers to be afraid of not meeting the taste parameters of their appellation rules if they were to reduce or dispense with sulphur inputs. In other words, the AOC/AOP committees (and their counterparts in other EU countries) might sometimes think that a wine with little or no sulphur was atypical of the appellation. And these fears are not entirely groundless. There are instances of producers not using sulphur being turned down because their wines do not meet the appellation's requirements of typicity, though this is not necessarily to say that the absence of sulphur is the root cause.

## What is typical?

This is our cue for a very interesting discussion. What exactly typifies an appellation? And does adding less sulphur make a wine less 'typical'? If by typical we mean what we are today accustomed to from traditional production, this is not infrequently the case. We have seen many such instances. Sometimes it is harder to recognise the origin of a wine with no sulphur. As Maddalena Pasqua at Musella, Valpolicella, puts it: 'We're in the process of converting to organic growing, but that doesn't mean we're going to stop using sulphur. We don't want to alter the style of our wine.'

Certain additives/processes which help to stabilise the wine can make it possible to reduce the sulphur input, for instance, flash pasteurisation or the addition of lysozymes, enzymes which stabilise must with a high pH. But many of the private organic and biodynamic labels prohibit both flash pasteurisation and lysozymes, and flash pasteurisation has the drawback of being an energy-guzzling process.

## Organic and authentic?

What makes a wine organic? Is it just the absence of pesticides and herbicides? Deciding which products may or may not be used out in the vineyard is relatively straightforward, but how do you go about deciding how a wine is to be made in the cellar? Do your indoor activities have any impact whatsoever on the outdoor environment? That is, if you do not go to the extreme of classing every vinification step requiring energy use as environmentally harmful.

But perhaps the point is not just the environment but also the imperative of an organic wine being as 'natural' as possible, and accordingly, that the smallest possible amount of additives should be used and the handling processes during vinification and storage kept as careful and few in number as possible.

ORWINE canvassed general public opinion on the subject of organic wine, and the fact is that one constantly recurring sentiment was that an organic wine must be 'authentic'. The study showed that people did not know very much about how a wine is made and what additives are used in the wine cellar. The consumers consulted wanted to prohibit, not only additives and operations harmful to human health but also everything that could affect the taste of the wine and, accordingly, render it less 'natural' and 'authentic'.

Additives which still have to be allowed – sulphur, for example – should, the respondents felt, be used in smaller quantities than for conventional wines. Additives deemed quite harmless included, for example, cultivated yeast, while opinions differed concerning oak chips, which were certainly rated a natural product, not a health hazard, but were felt to have a potential for spoiling the wine's authenticity. Moreover, people saw a risk of oak chips setting a precedent for the use of flavouring agents in organic wines. Anyone believing that must surely regard oak chips as a kind of flavouring agent, too, which is actually quite correct. Oak chips add taste to the wine without having any of the other effects on it that oak-barrel ageing does. So they are a questionable element in wines purporting to be authentic and a genuine expression of their origin, and many private labels forbid them. They are, however, sanctioned both by the National Organic Program and by Demeter USA.

So this debate is neither simple nor straightforward. Many of the techniques permitted by organic and biodynamic growers have, similarly, a direct impact on taste and could thus be seen as contrary to the notion of an 'authentic' or natural taste. Take oak-barrel ageing for example. This distinctly affects the taste of the wine. Add to this the effect on the wine of the degree of toasting, and you are not far from the same situation as with oak chips. But few, if any, put a case for banning the ageing of organic wines in oak barrels. Temperature-controlled fermentation is another example. Wine can turn radically different, depending on the fermentation temperature chosen (and artificially controlled). But no one is calling for a temperature control prohibition, either.

# CHAPTER 13

# Additives

There are today some eighty different additives which are permissible during the wine production process. There are products for stabilising, fining (clarifying), filtering, preserving, aromatising, acidifying and de-acidifying. There are enzymes and products which help the yeast during fermentation, preserve the colour, make the wine softer and so on.

By definition, an organic grower will also want to cut down on the use of additives, just as they scale down their spraying operations in the vineyard. The question, though, is what should be prohibited. Products that are harmful to the environment, of course. But all the things – additives and procedures – which do not harm the environment or humans, should they remain permissible? Should it simply be left to the producer to decide how far they want to go in their efforts to be 'natural'?

We find organic producers in every price bracket. There is a big difference between being a small, exclusive organic grower in Burgundy, selling one's wines for €50 or more a bottle, and an organic grower in southern Italy selling his for €2 per bottle. Both of them follow the rules applying in the vineyard, but obviously they cannot both work the same way, be it in the vineyard or inside the cellar. They are catering to different markets. A plain, mass-produced wine is not made the same way as a high-class, hand-crafted one. Flash pasteurisation can be termed acceptable for a cheap wine but not for a high-class, expensive *terroir* wine. And today both the top-end producer and the bulk producer can be organic.

*Vacuum distillation machine, Weingut, Schwarz, Austria.*

## Are additives and interference needed?

What additives are really necessary in wine production? Strict organic labels like Nature & Progrès and biodynamic Demeter permit, or at any rate tolerate, the following:

- 🌿 a small amount of sugar for chaptalisation.
- 🌿 bentonite and albumin for fining.
- 🌿 cellulose filtration.
- 🌿 temperature control of fermentation.
- 🌿 cold stabilisation of white wines.
- 🌿 sulphur – though in smaller quantities than indicated by general EU rules.

All this can, of course, be dispensed with and the wine made with absolutely no additives or interference, but what sort of wine will that be? Hardly anyone, no matter how natural they profess to be, would want to work like that. Virtually all producers and consumers feel that wine is actually improved by the use of certain additives and modifications inside the cellar.

A host of different additives and techniques are permitted and used. The quantity may seem horrifying, but there is really nothing peculiar about the products. The food industry employs them assiduously. One can be horrified by the additives present in food products as well, but that is another story …

The organic grower has to decide which additives/modifications are considered acceptable in an organic wine and must then make sure that these products, as far as possible, are of organic origin. Modifications should not be too energy-demanding, and the same goes for the production of the additives.

## Allergenic additives

All packaged foods must be accompanied by a package statement of any allergenic ingredients. Wines have been exempt from that rule, but at

*Sulphur can be administered the instant the grapes come in from harvesting.*

the beginning of 2012, following a recommendation by the European Food Safety Agency, the EU resolved that if a wine contains traces of egg or dairy products such as casein (milk protein), albumin (egg white) or lysozymes (an enzyme extracted from egg whites), then this must be stated on the label. The new rule came into force on July 1, 2012.

Albumin and casein are used for fining wines, and in most cases nothing remains of them after the fining, but traces can linger. Casein and albumin are used nowadays by many organic producers, but if so they are derived from organic livestock and eggs.

Otherwise all that has to be shown on a bottle of wine is the description 'Contains sulphur', meaning that the wine contains more than 10 mg sulphur per litre.

In Canada, too, a law entered into force in August 2012, requiring wine producers to include in their labelling a statement of all known allergenic products of which traces may be present in the wine.

## Which additives are used, and why?

So what are the additives used in wine-making? Which labels allow what? We will now turn to look at the different stages of vinification and the products and processes which the EU rules allow conventional growers to use, as well as a selection of organic labels and what they permit. The codes of practice we have included are as follows:

- The EU rule for all wines up to and including the 2011 vintage; as from 2012, these rules apply to conventional wines ('EU') only.
- The EU's new rules for organic wine, effective from 2012 ('EU organic').
- Demeter (biodynamic label).
- Nature & Progrès (N & P).
- Fédération Nationale Interprofessionnelle des Vins de l'Agriculture Biologique (FNIVAB).
- US Department of Agriculture (USDA), National Organic Program – Wine made with organic grapes (NOP).
- Bio Suisse.

# Additives and processes during vinification

*(Note that this is an incomplete list, giving only the commonest additives and processes.)*

## Protecting the newly picked grapes

Throughout vinification, the grapes/must/wine may need protection from excessive oxidation. The newly harvested grapes can be protected with ascorbic acid, sulphur dioxide and other sulphur compounds such as potassium bisulphite, ammonium bisulphite or potassium metabisulphite. All of these prevent oxidation.

### Sulphur dioxide ($SO_2$)
*Permitted by: all.*
*Forbidden by: none.*

### Ascorbic acid (Vitamin C, synthetically produced)
Also reduces heavy concentrations of iron or copper in the wine. One should beware of excessive dosage, which can give the wine a touch of bitterness. Makes possible a reduction of sulphur content.
*Permitted by: EU, EU organic, FNIVAB.*
*Forbidden by: Bio Suisse, Demeter and N & P.*

### Potassium bisulphite
*Permitted by: EU, EU organic, FNIVAB.*
*Forbidden by: NOP, Bio Suisse, Demeter and N & P.*

### Ammonium bisulphite
*Permitted by: EU, FNIVAB (certain types of wine).*
*Forbidden by: EU organic, NOP, Bio Suisse, Demeter and N & P.*

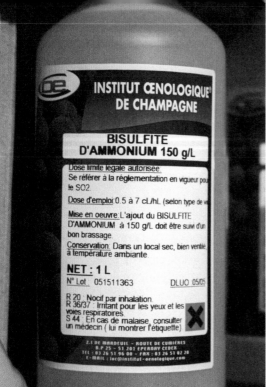

INSTITUT ŒNOLOGIQUE®
DE CHAMPAGNE

BISULFITE
D'AMMONIUM 150 g/L

Dose limite légale autorisée:
Se référer à la réglementation en vigueur pour
le SO2.
Dose d'emploi 0.5 à 7 cL/hL (selon type de vin)
Mise en oeuvre: L'ajout du BISULFITE
D'AMMONIUM à 150 g/L doit être suivi d'un
bon brassage.
Conservation: Dans un local sec, bien ventilé,
à température ambiante.

NET: 1 L
N° Lot: 051511363          DLUO: 05/05

R 20: Nocif par inhalation.
R 36/37: Irritant pour les yeux et les
voies respiratoires.
S 44: En cas de malaise, consulter
un médecin (lui montrer l'étiquette)

Z.I DE MARDEUIL – ROUTE DE CUMIÈRES
B.P 25 – 51 201 ÉPERNAY CEDEX
TÉL: 03 26 51 96 00 – FAX: 03 26 51 02 28
E-MAIL: ioc@institut-oenologique.com

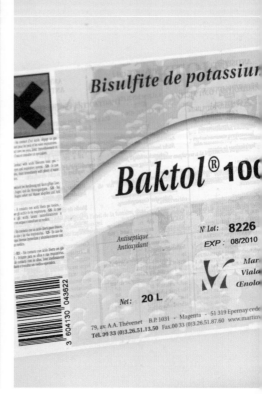

Bisulfite de potassium

Baktol® 100

Antiseptique
Antioxydant

N° Lot: 8226
EXP: 08/2010

Mar
Viala
Œnolog

Net: 20 L

79, av. A.A. Thévenet  B.P. 1031 – Magenta – 51 319 Epernay cede
Tél. 09 33 (0)3.26.51.13.50  Fax.00 33 (0)3.26.51.87.60  www.martin

3 604130 043622

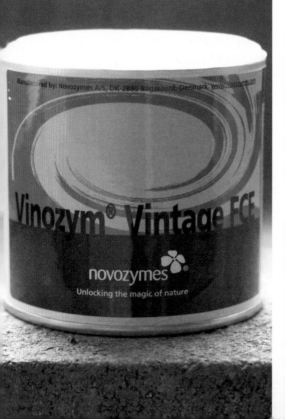

Manufactured by: Novozymes A/S, DK-2880 Bagsvaerd, Denmark. www.novozymes.com

Vinozym® Vintage FCE

novozymes®

Unlocking the magic of nature

RAPIDASE® CX

Enzyme
à activité
cinnamyl-estérase
naturellement
faible

naturellement FC

tydsa

ACIDO TARTÁRI
TARTARIC ACID BF

### Potassium metabisulphite

*Permitted by: EU, EU organic, Bio Suisse (certain types of wine).*
*Forbidden by: FNIVAB, NOP, Demeter and N & P.*
The newly picked grapes can also be protected with nitrogen gas or carbon dioxide. All labels allow this.

## In the fermentation tank

### Temperature control of the fermentation tank
Heating or cooling to the fermentation temperature desired.
*Permitted by: all.*
*Forbidden by: none.*

## Polyphenol extraction

### Addition of enzymes
Enzymes can be added to facilitate the extraction by the must of the important polyphenols (tannins and anthocyanins) present in the skins and seeds of the grapes. The enzymes may be pectines, cellulases or glycosidases. They are proteins derived from cultivated fungal organisms and they catalyse chemical reactions.
*Permitted by: EU, EU organic, FNIVAB and NOP.*
*Forbidden by: Bio Suisse, Demeter and N & P.*

## Getting the fermentation started

The following four products help fermentation, not only to get started but also to be completed and leave no residual sugar. This can be a problem if, for example, the must has a nitrogen deficiency or is cold, or if rot-infected grapes have been present in the tank.

### Thiamine (Vitamin B1)
Yeast nutrient: synthetically produced.
*Permitted by: EU, EU organic.*
*Forbidden by: FNIVAB, NOP, Bio Suisse, Demeter and N & P.*

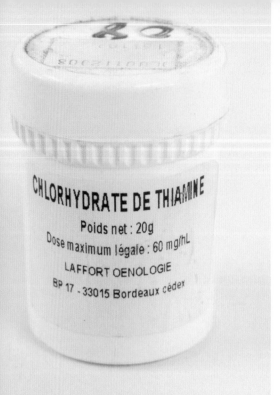
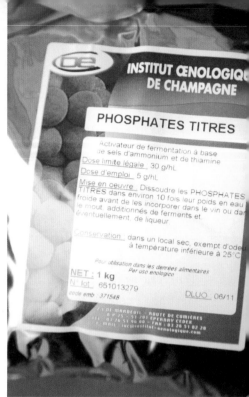

CHLORHYDRATE DE THIAMINE
Poids net : 20g
Dose maximum légale : 60 mg/hL
LAFFORT OENOLOGIE
BP 17 - 33015 Bordeaux cédex

INSTITUT ŒNOLOGIQUE
DE CHAMPAGNE

PHOSPHATES TITRES

Activateur de fermentation à base
de sels d'ammonium et de thiamine
Dose limite légale  30 g/hL
Dose d'emploi  5 g/hL
Mise en oeuvre  Dissoudre les PHOSPHATES
TITRES dans environ 10 fois leur poids en eau
froide avant de les incorporer dans le vin ou dans
le moût. additionnés de ferments et
éventuellement. de liqueur

Conservation  dans un local sec, exempt d'odeur
à température inférieure à 25°C

Pour utilisation dans les denrées alimentaires
Per uso enologico

NET : 1 kg
N° lot  651013279
code emb  371548
DLUO  06/11

### Yeast cell walls

Obtained from the yeast species Saccharomyses cerevisiae. Also helps
to stabilise the colour of red wines. The drawback is that the yeast cell
walls can make the wine smell of yeast.
*Permitted by: EU, EU organic, NOP.*
*Forbidden by: FNIVAB, Bio Suisse, Demeter and N & P.*

### Ammonium sulphate

*Permitted by: EU, Bio Suisse (subject to restrictions).*
*Forbidden by: EU organic, FNIVAB, NOP, Demeter and N & P.*

### Diammonium phosphate

*Permitted by: EU, EU organic and FNIVAB.*
*Forbidden by: NOP, Bio Suisse, Demeter and N & P.*

 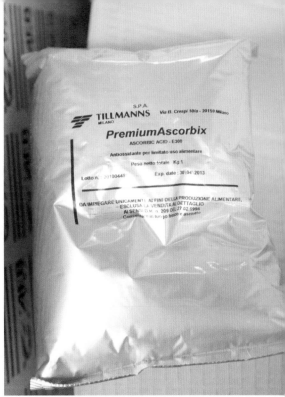

## Added or natural yeast?

Consulting oenologists often recommend playing safe and adding cultured yeast, arguing that relying on the native yeast naturally present in the grape skins and in the wine cellar is too chancy. The ecological associations beg to differ. If the grower is not using synthetic products that kill the wild yeast in the vineyard, natural yeast works excellently, in the opinion of Demeter and Nature & Progrès, who will only allow cultured yeast if you have very great difficulty in getting the fermentation started. Cultured yeast may not be used systematically. A wine with yeast added, in their view, will not be typical of its *terroir*.

There can be times when reliance on natural yeast comes hard: a white wine after one night's sedimentation or a white wine which also has to ferment at a very low temperature. It may also be that the wild yeast flora in the vineyard or the wine cellar is not so good. For organic growers, organic yeast is now commercially available.

ORWINE's studies show that roughly 60 per cent of organic growers in Europe add cultured yeast to their wine. The figure varies

somewhat from one country to another. In France it is only 51 per cent, and in Germany, Italy and Switzerland it ranges between 80-90.

### Cultured yeast

*Permitted by: EU, EU organic, FNIVAB, NOP, Bio Suisse.*
*Forbidden by: Demeter and N & P, but can be tolerated in certain circumstances, for example, stuck fermentation, and then only using yeast which is of organic origin and guaranteed not to be of genetically modified origin.*

## Raising the alcohol content – concentrating the must

Some years perhaps the must will not have sufficient sugar content to attain the alcohol strength desired in the finished wine. There are several different methods for raising the must's sugar content and with it the alcohol content of the finished wine. Chaptalisation, that is, the addition of sugar or concentrated must, is one of them. Sugar content can also be raised by drawing off some of the water from the must. This is done by vacuum distillation or reverse osmosis.

*Cultured yeast suitable for red wines.*  *Instructions for use on dry yeast packaging.*

Of course, the need to raise the alcohol content by artificial means is more commonly experienced in cool climates, but the same thing happens from time to time in districts like Bordeaux and the south of France. The organic labels are rightly sceptical of the practice, and the situation of no two vintages being alike is part and parcel of the organic and biodynamic mindset. The process, however, is allowed by several, though to a lesser extent than sanctioned by the general rules of the EU.

### Using ordinary sugar (beet or cane sugar)
*Permitted by:*

*EU: the alcohol strength may be raised by, at most, 1.5, 2 or 3 per cent, depending on whether you are located in a northern or southern zone.*

*EU organic: the same concentrations as the EU. 95 per cent organic sugar.*

*Bio Suisse: the alcohol strength may not be raised by more than 1.25 per cent.*

*NOP: no restrictions compared with conventional wines. Not more than*

*30 per cent of the sugar may be non-organic.*
*FNIVAB: the same levels as the EU. Only organic sugar allowed.*
*Demeter: only permissible for dry wines, and the alcohol strength may not be raised by more than 0.9 per cent. Only organic sugar allowed.*
*N & P: the alcohol strength may not be raised by more than 1 per cent. Only organic sugar allowed.*
*Forbidden by: none.*

## With concentrated grape must

*Permitted by: EU, EU organic, Bio Suisse, NOP, FNIVAB, N & P (all with the same restrictions on levels as stated above concerning ordinary sugar, all organic labels with organic must).*
*Forbidden by: Demeter.*

## With rectified concentrated grape must (Fr. Moût Concentré Rectifié – MCR)

*Permitted by: EU, EU organic, Bio Suisse, NOP, FNIVAB, N & P (all with the same restrictions on levels as stated above concerning ordinary sugar, all organic labels with organic must).*
*EU organic can be expected to insert a clause on subsequent revaluation of this rule.*
*Forbidden by: Demeter, NOP.*

## With vacuum distillation

A certain proportion of the water in the wine is boiled off at low pressure (and, by the same token, low temperature). There is a risk of some aromas being lost in the process. Expensive equipment, added to which, you lose roughly 8 per cent of the must (and, accordingly, of the volume of finished wine) for every single additional degree of alcohol strength. All components in the must are concentrated, which can have either advantages or disadvantages where colour and tannins are concerned.
*Permitted by: EU, NOP, Bio Suisse, N & P.*
*Forbidden by: EU organic, FNIVAB, Demeter.*

### With reverse osmosis

Some of the water is forced out of the must through a semi-permeable membrane. An expensive method. Here again, one loses roughly 8 per cent of the must per extra degree of alcohol strength, and all must components are concentrated.

*Permitted by: EU, Bio Suisse, NOP, FNIVAB, EU organic.*
*Forbidden by: Demeter and N & P.*
*(EU Organic will review this rule in 2015.)*


### Cryo-extraction

This method involves chilling the grapes below freezing point. They are pressed when the water in them is partly frozen, and in this way a more sugary must is extracted.

*Permitted by: EU, N & P.*
*Forbidden by: FNIVAB, Demeter, EU organic.*

## Malolactic fermentation

This is the so-called second fermentation, in which the sharp malic acid in the wine is converted to lactic acid. To get a safe (and sometimes early and fast) Malolactic fermentation, one can add lactic acid bacteria to the wine to help it along. Malolactic fermentation affects the wine in several ways: it reduces the acidity, stabilises the colour and tannins, and makes the wine seem softer and more buttery. This is desirable for all red wines and for some white ones (for instance, oaked Chardonnay).

In the case of aromatic white wines, sometimes one wants to preserve the wine's acidulous character and definitely does not want it to end up buttery. In that case, 'malo' is avoided by adding a little sulphur to kill the bacteria which will otherwise trigger the process. These bacteria, then, are naturally present in the wine, and in most cases malolactic fermentation will occur spontaneously, but, unlike the situation where bacteria are added, one does not always know exactly when. The process can be swift, but it can also take months. And so some producers opt for adding lactic acid bacteria. There is just a slight risk of lactic acid bacteria inducing higher volatile acids.

But, if you want malolactic fermentation to take place in autumn, directly after the alcoholic fermentation, this often means having to warm up the tank or cellar. One way of reducing energy consumption in the vineyard, as practised, for example, by Château Lagrange in Saint Julien (Bordeaux), is to add lactic acid bacteria in the tank at the same time as one adds yeast ('co-inoculation'). Malolactic fermentation will then take place simultaneously with or immediately after the alcoholic fermentation.

### Addition of lactic acid bacteria
*Permitted by: EU, EU organic, FNIVAB, NOP, Bio Suisse.*
*Forbidden by: Demeter and N & P (permission can be requested in connection with major difficulties).*

## Acidification

If the must is felt to be low on acid, tartaric acid or malic acid can be added to it or an electro-membrane employed.

This question mainly arises in really hot climates, but it can also arise in France during outstandingly hot summers. If tartaric or malic acid is to be added, this should be done before or during fermentation.

### Tartaric acid (Fr. acide tartrique)
One possible side-effect if too much is added is the wine taking on a bitter or astringent taste.
*Permitted by: EU, EU organic, FNIVAB, NOP, Bio Suisse.*
*Forbidden by: Demeter and N & P.*

### Malic acid
*Permitted by: EU, EU organic, FNIVAB, NOP, Bio Suisse.*
*Forbidden by: Demeter and N & P.*

## Electrodialysis

A high pH (and, accordingly, excessively low acidity) may be due to a surplus of potassium in the wine. Electrodialysis reduces potassium content by extracting potassium ions through a selectively permeable ion exchange membrane (the same technology as for desalinating sea water. The advantages are that this method entails low energy cost and at the same time stabilises the wine's tartaric acid content. The drawbacks are that the equipment is expensive to buy and the method can cause some discoloration of the wine (more intensity in the colour of red wines, which some, admittedly, may consider an advantage). Permitted within the EU since 2004 but not in any of the organic labels.

*Permitted by: EU.*

*Forbidden by: EU organic, FNIVAB, NOP, Bio Suisse, Demeter and N & P.*

## Acidity reduction

Acidity can be reduced by adding potassium bicarbonate, calcium carbonate or potassium tartrate. These neutralise the acids, tartaric acid especially, in the wine.

### Potassium bicarbonate

*Permitted by: EU, EU organic, FNIVAB, and N & P.*

*Forbidden by: Demeter, NOP, Bio Suisse.*

### Calcium carbonate

*Permitted by: EU, EU organic, Bio Suisse and NOP.*

*Forbidden by: FNIVAB, Demeter and N & P.*

### Potassium tartrate

*Permitted by: EU, EU organic and NOP.*

*Forbidden by: FNIVAB, Bio Suisse Demeter and N & P.*

## Stabilising the wine

A wine has many stresses and strains to withstand before landing in the glass, e.g. long transport distances or big temperature fluctuations. What the wine producer wants most of all is for their wine to taste as they intended, even if it is being drunk in another hemisphere. Accordingly, they need to ensure that the wine is stable when it leaves the winery.

Stabilisation covers microbiological stabilisation, colour stabilisation, fining and filtration.

## Microbiological stabilisation

### Sulphur dioxide and potassium bisulphite

Microbiological stabilisation by stemming the reproduction of bacteria and yeast. Anti-oxidant effect as well.

### Sulphur dioxide

*Permitted by: all (subject to restrictions – see table below, p. 181).*
*Forbidden by: none.*

### Potassium bisulphite

*Permitted by: EU, EU organic*
*Forbidden by: FNIVAB, Bio Suisse, NOP, Demeter and N & P.*

### Chitosan

Produced from fungal organisms. Can help to remove, for instance, brettanomyces (a yeast that can cause unpleasant smells) and heavy metals such as iron and lead.
*Permitted by: EU.*
*Forbidden by: EU organic, FNIVAB, Bio Suisse, NOP, Demeter and N & P.*

## Flash pasteurisation

The same method as used for pasteurising milk. It is coming to be used more and more in wine-making. Flash pasteurisation makes the wine biologically stable by killing off all micro organisms in it. The process involves quickly raising the temperature of the wine to between 70 and 80 degrees Celsius and then lowering it with the same rapidity after 20 or 30 seconds. The advantage is that flash-pasteurised wine needs less sulphur, the disadvantage that the method consumes a lot of energy.

*Permitted by: EU, NOP, FNIVAB, N & P.*

*Forbidden by: EU organic, Demeter, Bio Suisse.*

*EU organic allows the wine to be heated to not more than 65°C.*

## Lysozymes

Lysozymes are enzymes extracted from the whites of eggs. They are also used in the food industry, but the EU does not allow them in organic foodstuffs. Lysozymes can help stabilise the must and wine microbiologically by controlling the growth of lactic acid bacteria, which are 'responsible' for malolactic fermentation. They can delay and inhibit malolactic fermentation, thus constituting an alternative to sulphur as an inhibitor.

*Permitted by: EU, NOP.*

*Forbidden by: EU organic, Demeter, N & P, FNIVAB, Bio Suisse.*

## Dimethyl dicarbonate (DMDC)

A synthetic preservative or stabiliser which can be added to wines with sugar residue, so as to stabilise them microbiologically.

*Permitted by: EU (only wines with a residual sugar content exceeding 5 g/ litre. Dosage is heavily regulated, and the wine must not contain any traces of DMDC when released for sale.).*

*Forbidden by: EU organic, Demeter, N & P, FNIVAB, Bio Suisse, NOP.*

### Sorbic acid

A preservative which inhibits the growth of mould fungi, among other things.
*Permitted by: EU (in the form of potassium sorbate).*
*Forbidden by: EU organic, NOP, Demeter, N & P, FNIVAB, Bio Suisse.*

## Colour stabilisation

### PVPP

Polyvinyl polypyrrolidone (PVPP), a finely ground plastic product, is a synthetic fining agent of plastic origin which protects white wine must and rosé wine must from oxidation and stabilises their colour. It reduces the polyphenol content of the wine and removes bitter, astringent tannins. It is primarily used against a certain discoloration, but it can also be used for preserving the freshness and aromas of white wines. Known also by its E number, E1202.
*Permitted by: EU.*
*Forbidden by: EU organic, NOP, Demeter, FNIVAB, Bio Suisse and N & P.*

### Active carbon

Active carbon corrects the colour of white wines made from red grapes or white wine coloured by mistake or made from mouldy grapes. It has the effect of weakening the colour. It can reduce the intensity of the aromas.
*Permitted by: EU, EU organic, NOP, Demeter, FNIVAB, Bio Suisse (all subject to restrictions).*
*Forbidden by: N & P.*

### Casein

A milk product used mainly on white wines, as a colour corrective. Casein eliminates polyphenols which have oxidised or are likely to do so. It can also partly remove iron from white wine.
*Permitted by: EU, EU organic, FNIVAB, NOP, N & P.*
*Forbidden by: Bio Suisse, Demeter.*

*Sulphur-dosing equipment.*

 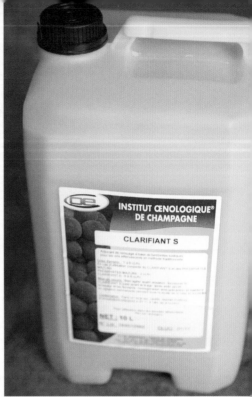

## Gum Arabic

A polysaccharide which improves the wine's stability in the bottle. Gum Arabic is a gum resin which comes from an African acacia species. It has two effects on wine, the first and most important being that it stabilises the colour, that is, prevents unwanted colour changes, above all if one wants the wine to retain a darker shade of colour. The second effect is that it can soften tannins and give the wine a more full-bodied character. Organic gum Arabic is commercially available.

*Permitted by: EU, EU organic, FNIVAB, NOP.*
*Forbidden by: Demeter, N & P.*

## Fining

Fining (clarification) means removing large molecules – colloids – which are unstable and can cloud or form sediment in the wine after bottling. Fining precedes filtration.

*Albumin (white of egg) for fining.*    *Cold stabilisation*

## Bentonite (a clay soil) common for white wines
*Permitted by: all.*
*Forbidden by: none.*


## Albumin – white of egg (red wines)
*Permitted by: all.*
*Forbidden by: none.*


## Gelatine
Extracted from the skins and bones of animals. Particularly useful with wines which are hard to clarify, for instance, because the grapes have been rot-infected. Softens red wines by removing some of their polyphenols.
*Permitted by: EU, EU organic, Bio Suisse.*
*Forbidden by: NOP, Demeter, FNIVAB and N & P.*

## Silica (silicon dioxide)

A natural mineral which has undergone chemical transformation. Facilitates fining.
*Permitted by: EU, EU organic, NOP, FNIVAB.*
*Forbidden by: Demeter, Bio Suisse, N & P.*

## Fish glue (Fr. Colle de poisson)

Used on white wines and some rosé ones. Made from the skin, fins and swim bladders of, mainly, sturgeon. Facilitates fining. Imparts a certain brilliance to white wine. Can remove certain polyphenols which would otherwise cause bitterness.
*Permitted by: EU, EU organic, Bio Suisse, FNIVAB.*
*Forbidden by: Demeter, N & P, NOP.*

## Filtration

### Kieselguhr

A fine powdery clay used for filtering.
*Permitted by: all.*
*Forbidden by: none.*

### Cellulose plates

*Permitted by: all.*
*Forbidden by: none.*

### Tangential filter

A membrane filter which does not clog up as easily as the ordinary kind.
*Permitted by: EU, EU organic (but only with pores larger than 0.2 micron), FNIVAB, Bio Suisse (only with pores larger than 0.2 micron), N & P. Demeter does not permit but tolerates.*
*Forbidden by: none.*

## Beta glucanase

The beta glucanase enzyme is used for facilitating filtration of botrytis-infected (sweet dessert) wines.
*Permitted by: EU, NOP, FNIVAB.*
*Forbidden by: EU organic, Demeter, N & P, Bio Suisse and FNIVAB.*


## Centrifuging

A rather uncommon method of removing particles from the wine. It can be performed with a greater or lesser degree of severity. It is uncertain whether EU organic is going to permit centrifuging.
*Permitted by: EU.*
*Forbidden by: Demeter, Bio Suisse.*


# Tartaric acid stabilisation


## Cold stabilisation

In time, the tartaric acid in the wine can combine with potassium to form potassium bitartrate, which at low temperatures can crystallise in the bottle and form what we call 'wine diamonds' or 'snowflakes'. This mainly occurs at low temperatures, for instance, if the wine is stored in a cool cellar or a refrigerator. The small crystals which become visible in the wine are perfectly natural but may have a negative impact on consumers. Some producers therefore prefer to ensure that the potassium bitartrate is already precipitated before bottling. Cold stabilisation entails chilling the wine to just over its freezing point (about −4°C) for roughly a week in the case of white wines and often a longer period for red ones. The potassium bitartrate can then be filtered off. The disadvantage of cold stabilisation is that it requires heavy energy inputs.
*Permitted by: all.*
*Forbidden by: none.*

## Metatartaric acid

Prevents precipitation of tartaric acid crystals. Adding metatartaric acid is an inexpensive method but with certain drawbacks. If the wine contains a lot of proteins, the metatartaric acid can cause a certain degree of clouding. The method's effect is of fairly brief duration, and so it is used mainly on wines for early drinking.

*Permitted by: EU, EU organic, NOP.*
*Forbidden by: Demeter, N & P, Bio Suisse, FNIVAB.*

## Electrodialysis

The advantages with this method (see above for details) are that it uses little energy and that both potassium and calcium bitartrates are precipitated, and that it gives reliable results. However, the equipment is expensive and is best suited for big wineries.

*Permitted by: EU, NOP.*
*Forbidden by: EU organic, Demeter, N & P, Bio Suisse and FNIVAB.*

## Ion exchange resin treatment

This improves the wine's stability. Potassium and calcium ions are replaced with hydrogen ions. The method can have the adverse effect of dulling the colour of the wine.

*Permitted by: EU.*
*Forbidden by: EU organic, NOP, Demeter, N & P, Bio Suisse and FNIVAB.*

## Mannoproteins

These improve the wine's stability by preventing precipitation of potassium bitartrate. Mannoprotein is a polysaccharide derived from yeast. Careful dosage is important, otherwise the wine may turn cloudy.

*Permitted by: EU, NOP.*
*Forbidden by: EU organic, Demeter, N & P, Bio Suisse and FNIVAB.*

# Other matters

## *Taste of reduction*

Reduction is the opposite of oxidation. A reduced wine can acquire a certain animal quality on the nose, but this usually disappears after a little aeration. In the case of too much taste (smell) of reduction, the smell can turn nasty, reminiscent of rotten eggs. The villain of the piece is hydrogen sulphide forming in the tank. The wine producer can eliminate the reductive taste by one or more rackings, but he can also remove it with the aid of copper citrate or copper sulphate.

The disadvantage of both copper citrate and copper sulphate is that they are liable to affect the wine's aromas.

### *Copper citrate*
*Permitted by: EU, EU organic.*
*Forbidden by: NOP, Demeter, N & P and FNIVAB.*

### *Copper sulphate*
*Permitted by: EU, EU organic (until 2015), Bio Suisse.*
*Forbidden by: NOP, Demeter, N & P and FNIVAB.*

### *Tannins*
Tannin can be added to the wine to improve the structure, stabilise the colour and remove the taste of reduction. Tannin can be derived from grape skins, pips, oak or chestnut.
*Permitted by: EU, EU organic, NOP, FNIVAB.*
*Forbidden by: Demeter, N & P, Bio Suisse.*

*Canned tannin.*

## Oak

### Oak barrels

No organic alternative is commercially available as yet, but certain forests are managed with sustainability certification from the Program for the Endorsement of Forest Certification Schemes (PEFC) or the Forest Stewardship Council (FSC).
*Permitted by: all.*
*Forbidden by: none.*

### Oak chips

These can be added during fermentation or else to the finished wine, either directly in the tank or in 'teabag' form.
*Permitted by: EU, EU organic, NOP, Demeter USA, FNIVAB.*
*Forbidden by: the rest of Demeter, N & P, Bio Suisse.*

# Sulphur

### Sulphur in red wines

| Sulphur addition, mg/l | Red wines, <2 g residual sugar/l | Red wines, <5 g residual sugar/l | Red wines, > 5 g residual sugar/l |
|---|---|---|---|
| EU rules for conventional wines | | 150 | 200 |
| EU rules for organic wines | 100 | 120 | 170 |
| Biodyvin | | 80 | 105 |
| Demeter International | | 110 | 140 |
| Nature & Progrès | | 70 | 130 |
| FNIVAB | | 100 | 150 |
| NOP | | 100 | 100 |

### Sulphur in white wines

| Sulphur addition, mg/l | White and rosé wines, <2 g residual sugar/l | White and rosé wines, <5 g residual sugar/l | White and rosé wines, > 5 g residual sugar/l |
|---|---|---|---|
| EU rules for conventional wines | | 200 | 250 |
| EU rules for organic wines | 150 | 170 | 220 |
| Biodyvin | | 105 | 130 |
| Demeter France | | 90 | 130 |
| Demeter International | | 140 | 180 |
| Nature & Progrès | | 90 | 130 |
| FNIVAB | | 120 | 130 |
| NOP | | 100 | 100 |

## Sulphur in sweet wines

| Sulphur addition, mg/l | Sweet wines (without botrytis)* | Sweet wines (with botrytis)* | Vin Doux Naturel |
|---|---|---|---|
| EU rules for conventional wines | 300 | 400 | 200 |
| EU rules for organic wines | 270 | 370 | 170 |
| Biodyvin | 175 | 200 | 100 |
| Demeter France | 130 | 200 | 80 |
| Demeter International | 250 | 360 | 140 |
| Nature & Progrès | 150 | 150 | 80 |
| FNIVAB | 250 | 360 | 80 |
| NOP | 100 | 100 | 100 |

\* The appellations and wine types concerned are defined in the EU regulations.

A beautiful oak-barrel cellar at Château Lagrange, Bordeaux.

# CHAPTER 14

# Sustainable Wine-Growing

Environmental friendliness in the vineyard and throughout the production and handling of wine is a hugely complex topic. Spraying or non-spraying of the vineyard is not all there is to it.

There are many to whom organic growing is not a realistic option. This is why so-called 'sustainable growing' has achieved such widespread impact all over the (wine) world. This approach is about conserving resources without jeopardising profitability, and so synthetic chemical products are permitted if necessary. But sustainable growers say the principle concerns not only spraying but also every other process in wine production capable of affecting the environment, such as power consumption, management of water resources, transport, and so on.

Sustainable and organic are not the same thing. One can be both, but one can also be sustainable without being organic, and vice versa (though, of course, a seriously organic grower should also embrace sustainability in a wider context than that of the vineyard).

Sustainable farming has developed in recent years along similar lines in various countries, and most wine-producing countries have now developed programmes for this type of wine production. Several also have control and certification systems.

## The definition of 'sustainable'

The UN defines sustainable development as 'development that meets the needs of the present without compromising the ability of future generations to meet their own needs.' (The Brundtland Report, 1987). The target for sustainable wine producers, then, is to carry on a

profitable business in that capacity while at the same time ensuring that they keep the soil healthy and sound. To them, profitability and organic are an equation that doesn't work out.

'Sustainable growing is the best thing,' say Jan and Caryl Panman of Château Rives-Blanques in Limoux in the south of France. 'That kind of farming is possible for most people, perhaps for everyone.' For a long time their estate was organic, but when it was hit by *flavescence dorée* they were forced to spray it with chemical pesticides or risk losing a large proportion of their vines. They have now been working at *raisonnée* for several years and are controlled by the official French sustainability label *Agriculture Raisonnée (AR)*. 'We follow Nature and have learned to react swiftly,' says Caryl Panman. 'We have learned to know our vineyard and our vines.'

There are of course risks involved in dispensing entirely with synthetic chemical spraying. Wet and damp conditions are a vineyard's worst enemy, and in certain climates and perhaps certain years, growers find justification for spraying so as not to lose part, possibly a large part, of the harvest. 'Showing responsibility but also taking account of an economic reality, because without viable estates there can be no sustainable wine

production,' according to Terra Vitis, one of France's private sustainability labelling associations. Sustainable growers feel unable to afford losing all or half their harvest to a disease in the vineyard. Many point out that in most years they are actually organic growers. The organic growers, of course, reject any such argument, insisting that you cannot be organic in certain years only. Either you are or you aren't. As they see it, if you occasionally resort to chemical products, your vineyard will never attain the equilibrium conferred by organic viticulture.

In recent years this group having sustainability as its objective has grown tremendously as an alternative to organic viticulture. But what exactly is meant by fair words like: 'Sustainable agriculture respects the environment, human beings and natural resources'?

## How do sustainable wine-growers do things?

'Sustainable' in French is often called *culture raisonnée* or *lutte raisonnée*, but also *culture durable*, but it is hard to grasp how 'reasonable' the sustainable growers are. The problem is that there is hardly anyone

*South African 'sustainable' wine
labelling, Groot Constantia.*

today who would frankly admit to using chemical sprays without
restraint. In that sense, everyone is *raisonnée*. The products used in
spraying are very expensive, and use is kept as low as possible, if not for
the environment's sake then for economic reasons.

Sustainability is a wide-ranging concept, and if a vineyard has no
certification it is hard to tell how they work. In fact, it is still hard to tell
even if they do have certification, because few products are expressly
banned. Instead the rules say that one must keep the use of pesticides
to a minimum. But of course, many growers take things seriously,
conforming to the guidelines laid down by the control agencies and
abiding by the principles of sustainability, in good faith and with
commendable ambition.

Detractors among the organic wine-growers will have us believe
that *agriculture raisonnée* is the brainchild of the agrochemical industry,
aimed at vindicating the use of its products. To the organic wine-
grower, *raisonnée* must always be a step on the way, with organic
growing, of necessity, as the ultimate destination. Most sustainable

*Flowers in between the rows of vines. Château Rives-Blanques, Limoux, Languedoc.*

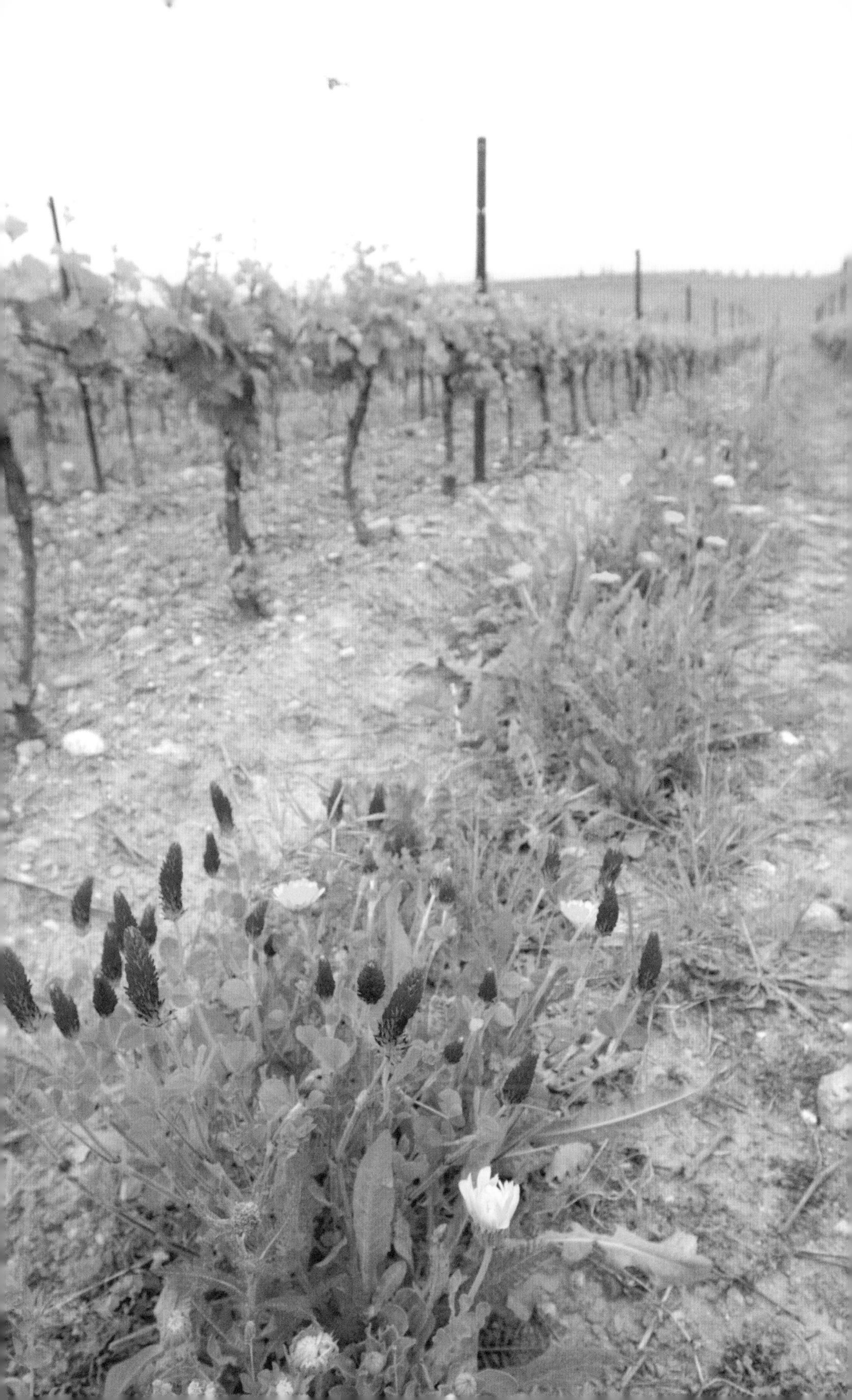

growers think nothing of the kind, feeling as they do that their view of the matter is more realistic, given that the market demands both quality and volume and that a vineyard is a business undertaking which has to be commercially viable.

By and large, the 'reasonables' work on the same lines as organic growers. They sow a cover crop in the vineyard, they take care to augment insect and plant life there, they plant hedges and so on. There are several possible ways of dealing with diseases and insect pests. Organic products are used, survival is made easier for natural enemies in the vineyard, sexual confusion is used, and weather stations to keep updated on the weather outlook, so as to avoid unnecessary spraying.

## Sustainability by the Cousinié Method

When Graham Nutter from the UK acquired Château Saint Jacques d'Alba in Minervois, Languedoc, he felt that working with respect for the environment, both in the vineyard itself and in its surroundings, was the natural thing to do. He chose to base his vineyard work on what is called *Méthode Cousinié,* a method created by Jean-Pierre Cousinié, using homeopathic products aimed at developing the vines' own defence system.

Graham describes how you begin by analysing the make-up of the soil. 'We analyse every patch of ground and then treat it according to need.' Any mineral deficiency is then rectified by spraying mineral powder on the leaves. In this way Graham has made up for zinc, potassium and nitrogen deficiencies in the soil. Synthetic products may be used if necessary. 'Better that than the disease spreading,' he says, but he regards copper spraying as a major problem which, moreover, can impede fermentation.

*Beehives can boost biodiversity. Domaine La Rune, Languedoc.*

## Labelling

Common to all codes of sustainability is their reference in general terms to the way in which the farmer/wine-grower is to tend the land and farm/estate. It is very seldom a particular product is banned. Instead the thing is to minimise negative environmental impact and to use, in small doses, the products which are least dangerous. Most tenets are recommendations. Tyflo in Alsace, with its restriction of the use of copper, makes an interesting exception. There they have decided that using less copper and supplementing it if necessary with other (synthetic chemical) products is better for the soil. This is an option which is not open to organic growers, who, in consequence, may need to use larger quantities of copper than the sustainable growers.

All labels attach great importance to traceability, that is, being able to trace what wine-growers have done throughout the production chain and which products they have made use of.

### France

'Agriculture Raisonnée' (AR) is an official French label controlled by the Ministry of Agriculture and administered by Farre (Forum de l'agriculture raisonnée respectueuse de l'environnement), an organisation marketing sustainable growing and registering growers who desire certification.

At present some one thousand vineyards have been awarded the label and are controlled by one or other of the nineteen accredited control agencies. Certification is awarded for five years at a time, and every fifth year the vineyard's soil is thoroughly inspected.

The rules are pretty vague. There are no direct prohibitions, and no chemical products are mentioned by name. It is more a matter of self-assessment. The producer must be motivated and assume personal responsibility for complying with the recommendations made concerning, for example, reduced spraying, priority for less hazardous products, inspection of the spraying equipment, reduction of spillage, and so on.

## Private associations in France

### Terra Vitis

Terra Vitis was first on the scene in France with organising *agriculture raisonnée* wine-growers. It started in Beaujolais in 1998 and now has members in Beaujolais, Burgundy, Languedoc-Roussillon, the Loire Valley and Bordeaux. Its members are controlled by Certipaq.

Terra Vitis has begun taking more and more interest in the overall environmental picture – energy, management of water resources, ways of increasing the diversity of flora and fauna, preserving ecosystems and so on. Great importance is attached to traceability, so that a check can easily be kept on the products which producers are employing. The idea is for producers to be all the time developing and improving their work in the light of new research findings.

### Tyflo

Tyflo certifies sustainable wine producers in Alsace. It has some precise rules, plus numerous recommendations. Downy mildew may be combated with up to 3 kg copper per hectare annually, which is 3 kg below the EU rules, but can be supplemented with synthetic products.

Tyflo does not allow any synthetic product to be used for combating botrytis. Instead the grower must take preventive measures – reduce the yield, reduce nitrogen fertiliser inputs, check mildew attacks, ventilate the canopy, keep the grape worm under (biological) control.

Every vineyard must have at least 5 per cent of its acreage set aside for 'ecological compensation zones'. These can take the form of hedges, woodland and meadowland. Their purpose is to harbour and feed the animal life which is beneficial to the vine.

Certain weeds and other grasses can compete with the vine, in which case they may need to be removed, chemically or mechanically.

193

In the rows of vines one can either let the grass grow or else sow a temporary or permanent cover crop in each or every other row. This grass must be removed mechanically. Chemical herbicides are permitted underneath the vines only.

## Austria

### Öpul Integrierte Produktion
An Austrian project for supporting and marketing sustainable agriculture.

## California

### The Sustainable Winegrowing Program

The Sustainable Winegrowing Program in California is based on three Es: Environmentally Sound, Socially Equitable and Economically Feasible.

The programme was launched in 2010 by the California Sustainable Winegrowing Alliance (CSWA) after three years' work in partnership with Californian wine producers.

To obtain certification, one has to be audited on site by a third party (a control organisation) which verifies the vineyard's compliance with a number of guidelines and criteria: saving water and energy, keeping the soil healthy, reducing pesticide use, protecting wildlife, protecting the quality of air and water, taking care of employees, and working to achieve the best wine quality. The vineyard must be aiming for continuous improvement.

Before any certification can come into question, the vineyards have to evaluate their work. In response to 277 different questions (criteria) they have to indicate on a scale from one to four how their winery corresponds to the description. In order to be considered for certification, they must score two or more on 58 of these requirements.

Annual improvement plans have to be drawn up for certain key

areas, and wine producers must be able to demonstrate improvements after a certain time. The motto reads: 'Sustainability is a journey not a destination.'

As yet, producers are not allowed to include this certification on their labels, but they can use it in their marketing and on their websites. The CSWA is forging ahead and will be drawing up metrics on water and energy use, greenhouse gas emissions and so on.

## Chile

### Certified Sustainable Wine of Chile

Chile has good natural preconditions for organic growing. Everyone seems agreed on this. The climate is dry and fine, healthy grapes are easy to produce. The country is isolated and diseases cannot make it here so very easily.

The number of organic and biodynamic Chilean vineyards is growing year by year, but much remains to be done to make wine producers and consumers more conscious of what this really means. Many are now moving towards a sustainable wine production which is to be environmentally friendly but economically sustainable and socially fair. The wine is to be seen as part of a whole comprising the winery, its employees and the local community. Certified Sustainable Wine of Chile marking became official in January 2012, with the certification of fourteen wine producers. They can now include this designation in their labelling. The fourteen are Anakena, Arboleda, Caliterra, Casa Silva, Cremaschi Furlotti, Emiliana, Errázuriz, Montes, MontGras, Santa Cruz, Santa Ema, Santa Rita, Ventisquero and Vía Wines.

## South Africa

Sustainable wine production is referred to in South Africa as Integrated Production of Wine (IPW). This label on a wine shows the consumer that it has been produced in a responsible manner. IPW was introduced by South Africa's wine industry in 1998, and the first vintage for which this certification could be seen on bottles from certain estates came in 2000. Certification is granted by the Wine and Spirit Board (WSB).

197

The guidelines are exhaustively detailed and refer to three stages: viticulture, cellar work and bottling. They are based on a point-scoring system. The estate fills in an annual questionnaire form about its working procedures and is awarded points depending on how environmentally friendly and responsible those procedures are. Independent auditors visit vineyards on a random basis every three years.

A combined label, Wine of Origin (WO) and IPW, has existed since the 2010 vintage.

Points of assessment for IPW marking are the following:

- how often spraying takes place.
- the spray used.
- how hazardous the products are.
- how they are stored.
- how pests are managed and combated.
- employee health and safety.
- waste management.
- irrigation.
- provisions on the planting of new land.
- rootstock selection.

Biodiversity in Wine Initiative (BWI) is another important marking system in South Africa, introduced at the initiative of the Botanical Society of South Africa as a way of preserving unique natural habitats in the South African wine-growing regions.

## New Zealand

### Sustainable Wine-growing New Zealand (SWNZ)

SWNZ's aim was for the whole of New Zealand's wine industry to be sustainable in 2012. That goal was almost but not quite achieved since at that date, according to their own report, 94 per cent of the country's wine producers were in the programme. It is also part of the plan for NZSW certification, in due course, to be a precondition for producers to export their wines.

# Nature

DOMAINE
La Fourmente

VISAN
CÔTES DU RHÔNE VILLAGES

ÉLEVÉ ET MIS EN BOUTEILLE AU DOMAINE
PAR LA FAMILLE POUIZIN A VISAN

Some synthetic spraying is permitted, but the idea is for it to be reduced, and insecticide spraying has already been cut by over 70 per cent, fungicide spraying by upwards of 60 per cent. But much work remains to be done to make consumers understand what the SWNZ label really stands for.

## Australia

### Sustaining Success
This is the name of a strategy for the whole of Australia's wine industry, and a label in the process of developing. Water quality is priority number one. Large amounts of water are used for irrigating Australia's vineyards. Large parts of the cultivated acreage are practically desert. All this irrigation brings problems, such as soil erosion and a steady growth of soil salinity.

### Generational Farming Program in McLaren Vale
The McLaren Vale wine-growing district of Victoria aims to become 100% self-sufficient in irrigation water. The McLaren Vale Water Plan helps growers to change for irrigation purposes from tap water to recycled water, that is, stormwater, mountain water and waste water which has been filtered and processed. 'In five years,' says Paul Rogers at the Fox Creek vineyard, 'we have raised reclaimed water use to 70 per cent of our irrigation water consumption.'

## Fairtrade

Social justice is part of the sustainability concept. This applies to all countries, but of course, the main focus of attention is on working conditions in Chile, Argentina and South Africa as wine-producing countries. Fairtrade is working with these three countries, and wine producers there can apply for Fairtrade marking to show that they care about this aspect. The Fairtrade label can be described as a kind of sustainability mark.

Fairtrade checks which chemicals are used in the vineyard and how, what working conditions are like, rates of pay, the minimum kilo price paid to local farmers delivering grapes to the producers, and so forth. Fairtrade also requires the vineyard to earmark a certain amount of money to be placed at the employees' disposal and spent, for example, on local schools.

## Code of Conduct for Nordic monopolies

In 2012 Systembolaget (Sweden's alcohol retailing monopoly), Alko (Finland's) and the Norwegian wine monopoly introduced a joint Code of Conduct which will have to be respected in order for a wine to be eligible for sale in their stores.

Systembolaget has been discussing the matter for some time, and the difficulty has lain in working out a definition and in finding an organisation which can offer the requisite expertise. Systembolaget has chosen Business Social Compliance Initiative (BSCI), based in Brussels, an organisation for companies wanting to take social responsibility and promote better working conditions. Lena Rogeman, head of Corporate Social Responsibility (CSR) at Systembolaget, says: 'BSCI is primarily concerned with social issues but addresses environmental and safety issues as well.'

Social issues are not Systembolaget's sole concern. Sustainable agriculture also matters. 'All producers will be compelled to latch on to sustainability,' says Master of Wine Ulf Sjödin, Systembolaget's Category Manager. 'Districts with adverse weather conditions cannot go organic, but sustainability is something that consumers are going to insist on.'

# CHAPTER 15

# *Vins Nature* – Natural Wines

It was only a few years ago that people began talking about 'natural wines' (*vins nature* in French), and now, all of a sudden, the term has become a buzzword, nearly more so than organic wine. There are wine bars and wine shops in Paris selling nothing but 'natural wines'. There was some confusion at first as to what this all meant, and there still is. Above all, there are no watertight compartments between organic, biodynamic and natural practices. There are organic and biodynamic growers who are just as natural as the natural ones. Anyone can call themselves 'natural' – there is no official definition of the word. If I buy a Demeter wine or a Nature & Progrès wine, I know more about the way it is made than when buying a wine purporting to be 'natural'.

Many users of the term 'natural wines' do not seem clear as to what it is really supposed to mean. Presumably what they have in mind is a wine which has been made without additives and without interventions in the cellar.

## French natural wines

Anyone wishing to join L'Association des Vins Naturels in France must conform to certain guidelines. These, however, are not mandatory rules. Instead the intention is for people to be able to trust each other, and no audits are carried out.

L'Association des Vins Naturels represents fifty or more wine-growers in France and elsewhere, as well as a variety of restaurants

*Egg-shaped concrete fermentation and storage tanks are used, for example, by certain biodynamic producers, but also by others. Boekenhoutskloof, Franschhoek, South Africa.*

selling natural wines. The rules, then, are loose-knit. It is desirable that wine-producing members should have organic certification. Failing that, you sign a declaration of compliance with the organic rules of the EU. The wine must be made using fresh, ripe grapes which have been subjected to an absolute minimum of manipulation. The grapes have to be picked by hand and fermented with natural yeast. All additives are to be avoided, apart from a small amount of sulphur. In other words, no acidity adjustment, no chaptalisation, no micro-oxidation, no reverse osmosis, and preferably no fining or filtration (though these two last mentioned are tolerated).

'Manipulation' is a word readily used in connection with natural wines, but this is really misleading. The French verb *manipuler* covers everything done by the wine maker in the wine cellar and affecting the wine: additives, mechanical operations such as pumping, filtration, thermal conditioning such as temperature control, and so on. The word does not in itself have any negative value attached to it such as 'manipulate' may have in English.


## Zero sulphur?

The term 'natural wines' is very often used to denote wines to which presumably no sulphur has been added. There are natural wines containing absolutely no sulphur, but many of them contain small amounts. L'Association des Vins Naturels recommends not exceeding the following concentrations:

- 30 mg/l for red and sparkling wines.
- 40 mg/l for dry white wines.
- 80 mg/l for white wines with a residual sugar content exceeding 5 grams/litre.

These are low concentrations by EU standards, but this can be the difference between a stable wine and an unstable one, the difference between a too pronounced taste of natural manure ('barnyard') and a good taste.

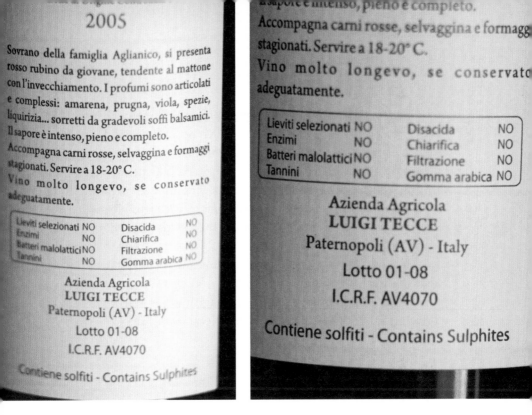

*An unusually detailed back label from Luigi Tecce, Campania.*

## Italian natural wines

Italy too has a strong natural wine movement. There are two Italian associations, one called ViniVeri (genuine wine) and the other VinNatur.

Both unite wine producers wishing to work with a minimum of intervention, both in the vineyard and in the cellar. Respect is to be shown for the *terroir* and natural cycles, and an effort made to produce individual wines. VinNatur runs courses for its members, collaborates with universities on research projects and holds an annual wine fair featuring natural wines from various countries. To verify members' compliance with natural methods, the association occasionally calls in bottles from them, and these are tested for traces of synthetic pesticides. VinNatur has about 140 members in nine different countries.

ViniVeri has eighteen members and very strict guidelines which have to be complied with. In addition to the ordinary rules of organic growing in the vineyard, growers have to plant local cultivars, harvest by hand and, when planting new vines, take cuttings from the vineyard's own vines (as opposed to going to the nursery and buying them). Only natural yeast may be used in the wine cellar, no additives are permitted, nor even temperature-controlled fermentation(!). Sulphur may be used, but not more than 80 mg/litre for dry wines and 100 mg for sweet ones.

Several members do not balk at experimenting and testing other types of material for fermenting the wine in. This has led to a small but increasing band of growers fermenting their wines in variously sized amphoras or clay jars dug into the ground. One of the first to do so was the now world-famous Jasko Gravner in Friuli, eastern Italy. Another Italian grower to do the same is Luigi Tecce in Taurasi, Campania, whose interesting amphora-fermented aglianico is warm and spicy, with a touch of honey (though of course completely dry).

## The taste

Well, what do these natural wines taste like? Sometimes they are quite 'normal'. They cannot be pigeonholed. In L'Association des Vins Naturels, for example, we have Jean-Pierre Frick in Alsace and Marcel Richaud in Cairanne, both very good at making wines with little or no sulphur and with no loss of freshness. Other people's natural wines, though, can feel prematurely oxidised. Sometimes the volatile acids and an exaggerated 'barnyard' nose get the upper hand. All wines contain volatile acids, some more than others, and some of the 'natural' wines come in the group of wines with high levels. Another reason may be that the natural mode of production exposes the wine to a great deal of oxygen, added to which, very little or no sulphur is used. If the wine has a high content of volatile acids, it can have been infected with acetobacter and begun slowly turning into vinegar. Sulphur fends off acetobacter, so there can undeniably be strong reasons for using a little.

Drinking wines with no sulphur (and with no other additives) is an interesting experience. Sometimes they can also taste good, but the taste is sometimes an acquired one. Above all, one needs to know the reason for these wines tasting differently. The natural-wine enthusiasts maintain that present-day wine consumers are too narrow-minded, that they all have a taste which is formed in the same mould and are therefore incapable of appreciating the finer points of natural wines.

One might think that a natural wine which is cloudy or oxidised is defective, but the enthusiasts maintain, on the contrary, that the producer wants it to look/taste that way. Or rather, Nature does.

There are many who profess and call themselves natural and many who are natural without proclaiming themselves as such. And there are degrees of 'naturality'. In the absence of a common or commonly accepted definition of the term, this is bound to be the case, and so one hesitates to make recommendations. But in Appendix 1 we give a list of examples of producers who, in our opinion, work well and in a natural manner (not necessarily with organic certification. If the producer belongs to one of the associations mentioned above, this is indicated in brackets).

# CHAPTER 16

# The Environmental Big Picture

Everyone knows now that environmental awareness means more than cutting down on spraying in the vineyard. Environmental impact can be reduced at all stages of the production process, from field to retail customer. This is all part of the philosophy, be it organic, biodynamic or sustainable. But environment protection can be taken to different lengths. Some producers try to be carbon dioxide neutral and self-sufficient in energy, while others make do with committing themselves to greener transport, lighter bottles, and so on.

## Carbon dioxide emissions

In 2008 the CIVB (Conseil Interprofessionel du Vin de Bordeaux) carried out a carbon dioxide analysis of the entire Bordelais wine industry, to quantify the greenhouse gas emissions it entailed. 'The wine industry is not exempt from global warming,' said the then CIVB Chairman Georges Haushalter. 'During the past twenty years we have seen a shortening of the growth cycle in Bordeaux, with earlier budding, which augments the risk of frost attacks.' He points to risks of changes in plant and insect life, of new insect pests appearing and of natural enemies disappearing. Research is underway in Bordeaux to test the viability there of non-traditional grape varieties, so as to prepare for the contingency of a warmer growth season.

Carbon auditing of a winery or, as here, of an entire region, is not altogether easy. There is quite a considerable margin of error. But at least you get some grasp of the direct and indirect carbon emissions

*Extremely low planting density and original vines. Luigi Tecce, Campania.*

resulting from work in the vineyard and cellar and from bottling, storage and delivery to customers. Measurement makes it easier to decide on measures for reducing dependence on non-renewable energy sources.

Altogether, according to this study, the Bordelais wine industry releases 200,100 tonnes of carbon dioxide equivalents annually (with a margin of error of 79,500 tonnes). Of this, 43 per cent, that is, 89,000 tonnes, emanate from 'incoming materials', the three biggest of which are glass bottles, cartons and pesticides. The manufacture of glass bottles and production of chemical pesticides are energy-guzzling processes.

Transport of wine to customers generates 37,000 tonnes of carbon dioxide (18 per cent of the total amount). Two-thirds of transport operations are by road and the remainder are divided between air and sea.

Energy consumption in the vineyard and during vinification and storage generates 30,000 tonnes, personal transport 23,000 tonnes, and real estate and machinery 16,000 tonnes.

Plan Climat 2020 in Bordeaux is aimed at reducing emissions by 40,000 tonnes between now and 2020 and the long-term target is a 75 per cent emission reduction by 2050.

Medium-term improvements will be:

- lighter bottles – manufacturing process improvement.
- transport to customers – development of environmentally friendly options.
- less use of chemical pesticides.
- general energy conservation.

A carbon dioxide audit was already carried out in 2005 by Château Lagrange, a classed growth in St Julien with a vineyard of more than 100 hectares. By far the biggest item (41 per cent), rather surprisingly, was the employees' travel, namely, commuter journeys and business trips. Tractor diesel emissions were nearly as high (39 per cent). Concrete responses to this audit included the estate commencing a review of its machine line-up with a view to cutting down on fuel consumption, and the introduction in the wine cellar of a new technique, *co-inoculation*, meaning the simultaneous administration of yeast and lactic acid

GRAND VIN DE BORDEAUX

2008

## LES HAUTS DE TRINTAUDON

### HAUT-MÉDOC

VIGNOBLE
RESPONSABLE

ÉVALUÉ PAR AFNOR CERTIFICATION

bacteria to the fermentation tank, which, as a result, no longer requires extra heating to get the malolactic fermentation started. This method cuts the estates electricity consumption by 8 per cent. In addition, bottles were introduced which were 10 per cent lighter than the old ones.

## Wine transport

Wine transport to customers entails heavy carbon emissions, whether by freighter from the other side of the globe or by road through Europe. Other transport operations, such as dispatching oak barrels from France to Australia or California, have to be included in the calculation. Substituting rail for road transport is a good idea but not always feasible. On the other hand, one can switch to more fuel efficient goods vehicles, improve transport planning so as to cluster destinations together, dispense as far as possible with refrigerated transport, and so on.

The environmental impact of transport is being widely noted today, for instance, in Chile. Viña Ventisquero, acting through British Climatecare, is investing in various environmentally friendly projects to offset carbon emissions caused by wine transport operations. These projects, which can take the form of wood-fired stoves in Uganda, wind farms in Turkey, new trees in the Amazonas, confer widespread social benefits in addition to their environmental advantages.

### Bulk shipping instead

Shipping wine in bulk instead of in bottles is an environmentally friendly alternative. Some European importers have now gone in for bottling New World wines at home in Europe, nearer the end customer. The drawback, of course, is that this takes local jobs away from wine-producing countries, for instance, South Africa, where job opportunities are greatly needed. Yet another example of environmental issues not always being so simple and straightforward.

## Carbon-neutral wines

Some producers go further still, and a few years ago the first completely carbon-neutral producer appeared. More and more are joining in. South America's first, Chilean De Martini, appeared in 2006, followed in 2007 by the first Californian estate, the Parducci Family Farmed Winery in Mendocino County, and the first French one, Vignoble Lacombe in Bordeaux. The Backberg Winery in Stellenbosch was the first South African estate to join in.

There is of course a limit to the amount by which a vineyard can reduce its emissions. To be carbon-neutral one must, quite simply, offset the emissions one cannot avoid. As Rémi Lacombe of Vignobles Lacombe, Médoc, puts it: 'We are a modern business undertaking, we cannot step backwards, and so when we found that we could not reduce our emissions any further, we decided to make up for them instead.'

The Backsburg Winery in Stellenbosch became South Africa's first carbon-neutral wine producer. Proprietor Michael Back considers it only natural to do what he can to soften the effects of global warming.

'The problem is our responsibility,' he says, 'and we must actively find solutions. We cannot temporise and leave it to the next generation to solve, because the next generation is in fact our children. To me, a wine coming out of a light-weight or PET bottle is just as enjoyable as one from a heavy glass bottle.'

Paul Dolan, formerly wine maker at Fetzer/Bonterra, now runs his own Parducci winery in Mendocino County, both organically and, partly, biodynamically. Throughout his career he has displayed profound environmental commitment. Here are some of his initiatives at Parducci:

- *Green energy:* 100 per cent of the estate's energy comes from renewables. 30 per cent is solar energy from solar collector panels installed on the estate, and the remainder comes from bought-in wind power from energy providers.
- *Water recycling:* The estate recycles all water used in the cellar for washing oak barrels and tanks. The water is purified with the aid of filters and a system of constructed wetlands and ponds (where wildlife flourishes – also a part of the firm's environmental initiative). The purified wastewater is used for irrigation and frost protection.
- *Water use:* Soil moisture is measured in the vineyard and water consumption metered at various points: in the wine cellar, the tasting room, the bottling plant, the guest house, the garden, and so on This way, a 25 per cent reduction has been achieved. Several producers now testify to the efficacy of installing a water meter. This makes it easier to see how much water you are using and, consequently, to use less.

So there are a number of measures which are really quite simple. For instance, keeping the air conditioning turned off during cool periods of the year, as is the practice, for example, at Torres in Chile.

214

## Lighter bottles

Putting one's wine into lighter bottles is one way of reducing environmental impact, both during the production of the bottles and when transporting the wine. The challenge lies in making lighter bottles which do not look or feel cheap, at all events until consumers get used to the fact of wine bottles not needing to be heavy. Normally an empty bottle weighs just under 500 grams. Today there are glass bottles which are between 10 and 50 per cent lighter. Consumers are accustomed to top-end wines coming in heavier bottles, and one can accept a certain difference in weight between everyday wines and prestige wines made in small quantities, so long as we are spared the worst of the monster bottles, which weigh in empty at over a kilo.

A few years ago, the Champagne district launched a new bottle weighing 835 g instead of the standard 900, that is, nearly twice the weight of an ordinary wine bottle, but necessarily so, given the high pressure inside a champagne bottle. A 65 g reduction may not seem all that much, but it still means more bottles per lorry and, accordingly, fewer lorries on the road. And the important thing for Champagne's image is that you cannot see or sense that the bottle is a bit lighter. In the world of champagne, appearances count for a great deal!

The bigger the packaging's format, the more environmentally friendly it will, of course, be to transport and manufacture. Large BiB packagings score well in this respect, as do PET bottles (but will consumers accept wine in PET bottles?).

In 2010 Systembolaget (the Swedish monopoly retailer) carried out a study concerning the environmental impact of different packagings. The study showed Bag-in-Box to have less environmental impact than glass bottles. PET bottles came in between glass ones and cardboard packagings, but, on the other hand, are more environmentally friendly from a transport viewpoint because they weigh so little. The study was carried out in collaboration with Vinmonopolet Norge (the Norwegian monopoly retailer), Tetra Pak Sverige AB, Elopak GroupNorge, Oenoforos and Vitop Smurfit Kappa Bag-in-box. All of these, except for Oenoforos, a wine importer, and the two monopolies, are producers of Bag-in-Box and Tetra Pak (a cardboard packaging), and so perhaps the findings have to be interpreted keeping that in

Garantie Fraîcheur jusqu'à
**6 SEMAINES**
APRÈS OUVERTURE

WINE Pouch®

SYRAH GRENACHE
VIN ROSÉ

*Vin issu de raisins biologiques*

**[BIVOUAC]**

CERTIFIÉ
**AB**
AGRICULTURE
BIOLOGIQUE

VIN ISSU DE RAISINS
DE L'AGRICULTURE BIOLOGIQUE
CERTIFIÉ PAR ECOCERT FR-Bio-01

**80%**
d'empreinte carbone en moins

**90%**
de déchets en moins*

**100%**
de raisins issus de l'agriculture biologique

OSCAR
de l'emballage
LAURÉAT 2009

1,5L =

mind. Producers of other types of packaging were approached but declined to take part in the study.

Labelling too can be made environmentally friendly through the use of recycled paper and natural glue. Torres in Chile uses a natural glue made from casein and albumin.

## Natural cork

Ought one preferably to use natural cork or a screw cap? There is much to suggest that natural cork is the environmentally friendlier of the two.

In 2009 PriceWaterhouseCoopers compiled a report stating that carbon emissions during the lifecycle of a screw cap were 24 times those of a natural cork, and that those from a plastic cork were ten times higher than for a natural one. True, the report was commissioned by Amorim, Portugal's biggest cork producer, but the study was purportedly conducted with strict impartiality and in accordance with ISO 14040 and 14044. But, just as with the Systembolaget study mentioned above, there is no harm in knowing who the surveys were commissioned by.

Once the bark has been harvested from a cork oak there is no spillage. Everything can be used – the very sawdust is swept up and used as fuel in the factory furnaces. The corks can be recycled by grinding them down for incorporation in various products (not, however, including new wine corks). And even if they are not recycled, corks degrade without generating environmentally hazardous waste.

The World Wildlife Fund puts another slant on the environmental advantages of cork. A WWF report published in 2006 forecasts drastic consequences in the event of natural cork being discontinued on a large scale as a closure device. In the long term, this would probably spell the demise of the cork oak forests. Those forests have an astonishing diversity of plant life and provide habitats for several endangered animal species. Cork oak forests, the WWF argues, have a rarely seen diversity of plants, trees and wildlife which must on all accounts be preserved, and this diversity is created in forests tended and used by man.

*Bagged wine – an environmentally friendly packaging.*

217

Portugal's cork oak forests have more than 13,000 native plant species, the greatest number anywhere, we are told, next to the tropical Andes. There are 150 tree species, of which the cork oak is one. The forests are an important stopover for birds migrating southwards. The tree roots bind the soil, preventing erosion. The trees absorb carbon dioxide, and a cork oak regenerating its bark absorbs between three and five times more than one from which no bark is taken. The forests impede forest fires, because cork oaks do not burn as well as other trees. And, not least, even if it has little environmental impact, the acorns of the cork oak provide food for the little black pigs which run about in between them and are eventually turned into good dry cured ham.

The fantastic thing about the cork oak is that you can use it without having to fell it. You harvest the bark, and fresh bark then grows in its place. Bark can be harvested between fifteen and eighteen times in the tree's lifetime. Normally a tree will live for between 150 and 200 years.

If the steady growth of alternative closure devices continues, the WWF fears that sales of natural cork will plummet by half before 2020, in which case harvesting would be discontinued in a million hectares of cork oak forest (out of a world total of 2.2 million ha). True, natural cork is still being used for many other purposes but 70 per cent of the cork industry's revenues come from wine corks, and so without them people would not be so very interested in looking after the forests.

## Choosing the right oak forest

Proper care and sustainable management of forests means we can go on having a good supply of oak barrels and natural corks. A forest with Forest Stewardship Council (FSC) certification is manifestly being managed in exactly that way. The FSC is an independent body which inspects and certifies forests the world over.

## Solar collector panels

Solar collector panels are beginning to appear in all kinds of places in the world of wine – not only at Paul Dolan's – and sometimes in unexpectedly cool spots, such as Champagne Duval-Leroy or the Loire Valley in France.

Caves des Vignerons de Buzet in the southwest of France, a co-operative with a very active, dynamic environmental policy, recently gave its wine cellar a new roof, consisting of 8,000 square metres solar collector panels and constructed by a power company in return for access to the electricity which the roof generates, equalling the annual consumption of 185 French households.

Frédéric Brochet of Domaine Ampelidae in the southern Loire Valley dreamed of making his estate as self-sufficient as possible. That dream came partly true in 2009, when the local authority gave him a funding grant for the installation of 400 square metres of solar collector panels at Ampelidae. The northern location gets just enough sun for solar collectors to be worth while. 'We are now self-sufficient in electricity,' he says, though this is not quite the whole truth. The solar collectors do indeed generate as much electricity as the estate needs in a year, but, understandably enough, more is produced in summer, and so then he sells off the surplus, while in winter he has to buy energy. Storing electrical power would be far too complicated.

## Irrigation

In many countries, irrigation is unavoidable. There are districts in hot, arid countries which are wholly dependent on irrigation and would not exist without it. Irrigation in Australia has caused major environmental problems, and the water management issue today is a prime concern of the Australian environmentalist movement.

In South Africa too, water management is a major issue. Efforts are now being made to reinstate the local, original vegetation round the vineyards, which requires less water than, for example, imported trees like the eucalyptus.

## Trapping carbon dioxide

One inevitable effect of wine production is the formation of carbon dioxide during fermentation. Suppose that carbon dioxide could be captured and perhaps used for something? Xavier Planty, one of the owners of Château Guiraud, Sauternes, had that very idea and was able to translate it into practice when he joined forces with the French Alcion Environnement company. Alcion have now developed a technology which captures the carbon dioxide and converts it to bicarbonate. Ingenious? It certainly sounds like it.

The users include two co-operatives: la Cave des Vignerons de Tutiac in Bordeaux, and la Cave des Vignerons de Buzet. The Buzet co-operative is ready to employ the technique on a large scale, and there are just a few minor points of detail to be sorted out. Les Vignerons de Tutiac, in the southwest, used the method in some of their red-wine tanks for the 2011 vintage. 'We're still evaluating,' says Lydia Héraud, their marketing director, 'but we're optimistic.'

Jean Philippe Richard of Alcion sums up the benefits as follows: 'The environment in the fermentation room is made less hazardous, carbon emissions are reduced and, in addition, bicarbonate is a product with many different uses – in cooking, for household care products, in cleaning agents and more besides.' In purely economic terms, at present the technique is mainly of interest where there are really large volumes involved.

*Irrigation equipment at Raats Family Wines, South Africa.*

# APPENDIX 1

# Recommended 'natural wine' producers

*(Affiliation to relevant bodies shown in brackets.)*

## France

### Alsace:

Domaine Christian Binner (AVN)
Domaine Pierre Frick (AVN)
Domaine Julien Meyer (AVN)

### Burgundy:

Domaine Jean-Paul Thévenet (AVN)
Domaine Catherine et Dominique Derain

### Beaujolais:

Domaine des Terres Dorées (VinNatur)
Domaine Marcel Pierre
Domaine Jean Foillard

### Southwest:

Domaine Causse Marines, Gaillac
Domaine Richard, Bergerac

### Loire:

Clos du Tue-Boeuf, Touraine (AVN)
Olivier Cousin, Anjou (Vin Natur)
Pierre Breton, Bourgueil (AVN)
Domaine des Maisons Brûlées (AVN)
Sébastien Riffault, Sancerre (AVN och VinNatur)
Philippe Tessier, Cheverny (AVN)
La Ferme de la Sansonnière, Anjou

### Rhône:

Domaine Gramenon
Domaine de la Ferme Saint-Martin

*Vineyards with the Andes in the background in Región del Maule, Chile.*

Domaine Marcel Richaud
Domaine Viret

## *Languedoc:*

Domaine du Fontedicto (AVN)
Domaine Turner-Pageot
Domaine Léon Barral

## *Champagne:*

Leclert-Briant
Jacques Selosse
Fleury

## *Corsica:*

Antoine Arena (AVN)

---

# Italy

## *Friuli:*

Gravner
Zidarich (ViniVeri)
Radikon (ViniVeri)

## *Toscana:*

Colombaia

## *Piemonte:*

Teobaldo Cappellano (ViniVeri)
Trinchero (ViniVeri)

## *Campania:*

Luigi Tecce

## *Sicily:*

Frank Cornelissen (VinNatur and AVN)

## *Emilia-Romagna:*

Alberto Tedeschi (VinNatur)

---

# Portugal

Quinta do Infantado, Douro (VinNatur)

# APPENDIX 2

# A selection of favourite organic and biodynamic wine producers

This is a selection of some wine producers whose wines we have tasted and have liked. This is in no way a 'complete' list of recommended producers. That would be an impossible task to do. There are thousands of wine producers that are organic or biodynamic, many of them excellent, but we have not tasted them all. It is a segment of the market that is changing fast. New producers convert continuously to these methods of farming. So take this as suggestions for wines and producers to try, but do also find your own favourites outside this list. There are so many talented wine producers out there!

## France

### *Languedoc-Roussillon*

#### *Château Pech-Latt, Corbières*
One of the organic pioneers in Languedoc. Certified since 1991. The estate has one hundred hectares of vines and is quite isolated with no close neighbours. A big advantage, says director and wine maker Philippe Mathias.

#### *Domaine Turner-Pageot, Languedoc*
Organic and Demeter certified since 2012. 'Being biodynamic gives us better balance in the wine', says winemaker Emmanuel Pageot, 'and also a better acidity even though we have riper fruit'.

## Domaine Jean-Baptiste Senat, Minervois

Certified organic since 2009. Jean-Baptiste moved from Paris to Languedoc in 1995. 'The health of the vines is so important', he says, 'that converting to organic viticulture was natural for me'. He is particularly fond of Carignan which, he feels, adds freshness and structure to Grenache.

## Château Pech-Redon, Languedoc-La Clape

Certified organic since 2009. In the very beautiful region of La Clape, close to the Mediterranean, Christophe Bousquet makes affordable wines with personality and character. He is one of the pioneers for quality in the Languedoc. He stopped using synthetic products in 1998. The vines produces less grapes but that is good, he says: 'Chemical products forces the vines to produce more than they are capable of'.

## Domaine Sainte Croix, Fraïssé des Corbières, Corbières

Certified organic since 2009. Jon Bowen is English and had worked some years as a wine maker in the Languedoc already when he and his wife Elizabeth started their own business. They prefer fruity, easy drinking wines but still with complexity. Jon only uses natural yeast for fermentation.

## Domaine Olivier Pithon, Calce, Côtes de Roussillon

Certified organic since 2002. Olivier Pithon is lucky to have very old vines of Carignan, some of them planted in 1900. The schist and the sea breezes give his wines a freshness you would not expect in the hot region. He is very inspired by the biodynamic methods but does not want to follow all the rules.

## Domaine Gauby, Calce, Côtes de Roussillon

The first producer to discover the magnificent terroir of Calce. Domaine Gauby's natural way to work, with few additives and low sulphur content, puts its mark on the wine. A lot of Syrah and Mourvèdre gives body and richness. Plums, herbs and some 'animal' character.

*Cabernet franc grapes at the Domaine des Roches Neuves, Saumur-Champigny.*

### Domaine Cazes, Rivesaltes, Roussillon

Certified biodynamic through Biodyvin. Domaine Cazes in Rivesaltes shows that also big estates can work organic and biodynamic. Domaine Cazes have 190 hectares of vines which makes it one of the biggest biodynamic estates in Europe and the world. 'The most noticeable change', says Emmanuel Cazes, 'is that the wines are now much fresher. Fifteen years ago we had a lack of acidity in our wines. It never happens now'.

### Domaine Monplézy, Languedoc-Pézenas, Côtes de Thongue

Certified organic since 2012. The estate is beautifully situated and surrounded by the typical *garrigue* of the Languedoc with its fragrant herbs. The 22 hectares are run by Anne Sutra de Germa and her husband Christian Gil. Fruity, drinkable wines, the reds from Carignan, Grenache and Syrah and the pleasant whites from Marsanne, Roussanne and Viognier.

### Domaine Alain Chabanon, Montpeyroux

Certified organic since 2005. Alain Chabanon never sprays, he says, not anything. And he never adds any fertiliser to the soil, not even of natural origin. His natural thinking continues in the cellar. The must ferments without temperature control and, of course, only with the natural yeast. His wines are intense with concentrated aromas and complexity.

### Château Maris, Minervois

Englishman Robert Eden works with extremely low yields which makes for a lot of concentration in the wine. You find dark berries, density and texture in his wines but also good fresh acidity that balances the high alcohol contents.

### Domaine Léon Barral, Faugères

Didier Barral works as naturally as possible and he even has cows and horses in the vineyard.

### Domaine Borie La Vitarèle, Saint Chinian

Pure and fresh wines from a beautiful and very well protected vineyard up in the Saint Chinian mountains.

## Domaine Jean-Louis Denois, Limoux

Interesting wines from Pinot Noir, Chardonnay and Merlot from a very skilled winemaker. Some of them with no sulphur at all.

## Domaine le Conte des Floris, Languedoc-Pézenas

Daniel Le Conte des Floris makes brilliant, very intense white wines, often from Grenache blanc and Roussanne.

## Domaine Roc des Anges, Roussillon

Marjorie Gallet has her estate in the beautiful Vallée de l'Agly northeast of Perpignan. The complexity in her wines comes partly from very old vines, over one hundred years old, of Carignan and Grenache.

## Mas de Soleilla, Languedoc La Clape

Swiss winemaker Peter Wildbolz makes full-bodied, perfumed and quite elegant wines here at sun drenched La Clape close to Narbonne and the Mediterranean.

## Domaine d'Aupilhac, Montpeyroux

Sylvain Fadat is one of the leading lights in the small village of Montpeyroux. He makes some excellent wines that range from every-day drinking to very concentrated and ambitious *cuvées*. Some of his vineyards are spectacularly high located up in the Languedoc hills.

## Château de Cazeneuve, Pic Saint Loup

André Leenhardt's vineyards are on a plateau in the Pic St Loup region in the northern Languedoc. His wines are full-bodied and concentrated.

## Domaine Lacroix-Vanel, Languedoc

You find Jean-Pierre Vanel in the small, rural town of Caux where he is one of several talented winemakers. His wines carry names alluding to music and his wines are correspondingly elegant.

## Domaine Canet-Valette, Saint Chinian

Marc Valette always goes his own ways. He was a pioneer in the Languedoc quality revolution. Deliciously drinkable yet affordable wines.

### Domaine Piccinini, Minervois-La Livinière
Jean-Christophe Piccinini's father, Maurice, was for many years in charge of the village co-operative in La Livinière and turned it into one of the best in the region. Jean-Christophe launched out on his own and his wines are today among the best wines from Minervois.

### Château des Estanilles, Faugères
Château des Estanilles was for many years one of the best known estates in the Faugères. It changed hands a few years back and the new owner, Julien Seydoux, has taken it even further with a strong enthusiasm for quality and organic farming.

### Clos de l'Anhel, Corbières
A very small winery created and run by Sophie Guiraudon. She has no family history in winemaking and has created a winery that consistently makes very likeable wines with a lot of individuality.

## And some more recommendations from Languedoc:

Domaine Bertrand-Bergé, Fitou
Domaine Pierre Clavel, Pic Saint Loup
Domaine du Château de Caraguilhes, Corbières
Domaine Grand Guilhem, Corbières
Domaine Zumbaum Tomasi, Pic Saint Loup
Domaine Singla, Roussillon
Mas Bruguières, Pic Saint Loup
Domaine Ermitage du Pic Saint Loup, Pic Saint Loup
Château de Montpezat, Languedoc Pézenas
Mas de l'Ecriture, Jonquières, Languedoc
Domaine Pouderoux, Roussillon
Domaine Clos Marie, Pic Saint Loup
Domaine Grécaux, Montpeyroux
Mas Cal Demoura, Languedoc Terrasses du Larzac
Château Coupe-Roses, Minervois
Château Beauregard, Corbières
Domaine Boucabeille, Roussillon

*The dramatic mountain formations in Franschhoek, South Africa*

## Rhône Valley

### M. Chapoutier, Rhône Valley and Roussillon

Certified Biodyvin and Demeter since 2000. 'The soil needs to get its natural microbiological life back to enable the wine to have a terroir taste', says Michel Chapoutier. 'With biodynamics', he adds, 'you get an additional complexity in the wines, another dimension. If the wines are any better is up to the customer to decide'.

### Domaine Alain Voge, Cornas, Saint Péray

Organic since 2006. Wine maker Albéric Mazoyer used to be technical director at Chapoutier. The slopes here in Cornas can be quite steep. 'The biggest problem with being organic', says Albéric, 'is not downy mildew or grey rot, it is the grass growing in the vineyard. It is very time consuming to remove it without herbicides'.

### Domaine Duseigneur, Lirac, Côtes du Rhône-Villages-Laudun

Certified Biodyvin since 2002. Bernard Duseigneur is very quality conscious and aims at elegance and finesse in the wines. Decoctions from different herbs and plants, many growing around the winery, help keep the vines healthy and the soil in harmony.

### Domaine Montirius, Vacqueyras

Certified Biodyvin since 1999. Christine and Eric Saurel met biodynamic consultant François Bouchet in 1996 and it changed their lives. They were sceptical to start with but were soon convinced. 'We did not have any downy mildew in 1997 although all our neighbours were badly hit', says Christine as an example. She stresses one important thing: 'People have many false ideas about biodynamics. It is important to understand that we are first and foremost winemakers'.

### Domaine Le Sang des Cailloux, Vacqueyras

Certified Biodyvin since 2011. 'We have always spent a lot of time in the vineyard', says Serge Férigoule, the colourful owner of Domaine Le Sang des Cailloux, 'and we spend even more time there now, being biodynamic'. His philosophy in the cellar is to supervise and to interfere as little as possible. It is easy if you have worked well in the vineyard, he says. At least easier…

234

### Domaine La Cabotte, Côtes du Rhône

Certified Demeter since 2010. Marie-Pierre Plumet d'Ardhuy and her husband Eric Plumet make delicious wines in the heart of the beautiful Massif d'Uchaux. Marie-Pierre has been through all stages of viticulture, conventional, *lutte raisonnée,* organic and finally biodynamic. 'Now our wines have more personality', she says.

### Domaine Gourt de Mautens, Rasteau

Certified Demeter since 2008. For Jérôme Bressy being biodynamic has helped him to create harmony in the vineyard and a sense of *terroir* in his wines. Jérôme took over the family estate in 1996 and aimed for high quality to start with. He works with parcel selection, very low yields, long extractions, ripe grapes and no destemming.

### Domaine de la Monardière, Vacqueyras

Certified organic since 2007. Christian Vache, his wife Martine and son Damien feel that organic agriculture improves the quality of their wines. Canopy management combined with the persistent mistral wind keep the vineyard dry so treatments with sulphur and copper can be kept to a minimum.

### Domaine Saladin, Côtes du Rhône

Domaine Saladin is located in the northern part of the southern Rhône Valley. It is run by two young sisters: Elisabeth and Marie-Laurence Saladin, twentieth (!) generation winemaker. They make a range of wines that compare very favourably with many from more famous appellations, including one from a vineyard that was bought by the Saladins in 1422.

## And some more recommendations from the Rhône Valley

Domaine Clusel Roch, Côte Rôtie
Domaine Combier, Crozes-Hermitage
Domaine Ferraton, Hermitage/Crozes-Hermitage
Domaine de Cristia, Châteauneuf-du-Pape
Domaine Pierre André, Châteauneuf-du-Pape
Domaine La Fourmente, Visan

Domaine du Trapadis, Rasteau
Domaine Gramenon, Côtes-du-Rhône
Château La Nerthe, Châteauneuf-du-Pape
Domaine Marcoux, Châteauneuf-du-Pape
Domaine Gilles Robin, Crozes-Hermitage
Domaine Chave (Yann Chave), Hermitage
Domaine des Entrefaux, Crozes-Hermitage
Domaine de Fauterie, Saint-Péray
Clos des Papes, Châteauneuf-du-Pape

## Loire Valley

### Clos de la Coulée de Serrant, Savennières

Certified Demeter since 1984. No one promotes the natural and genuine wine more intensely than Nicolas Joly. Wine can, and should, be made without additives, according to this famous preacher for biodynamic agriculture. His estate was one of the first in France to convert. His daughter Virginie is now more and more involved in the wine making.

### Domaine des Roches Neuves, Saumur-Champigny

Certified organic Biodyvin since 2003. Owner and wine maker Thierry Germain looks for elegance, freshness and cleanness in the wines, something, he says, he gets with biodynamic agriculture. His wines are good examples of how delicious the Cabernet Franc grape can be. With grass flourishing in the vineyard the insects are back and nature is in balance. Thierry also points out that the grape bunches are less tight making them less susceptible to rot.

### Ampelidae

Certified organic since 2007. Frédéric Brochet is owner and winemaker at Ampelidae, an interesting estate half way between Loire and Bordeaux. The most important thing when you go organic is to stay vigilant, he says, especially during difficult and rainy years. Copper is a problem he thinks, something that needs to be dealt with. He guarantees that his wines are a 100 per cent free from any pesticide

*Terraced vineyards in Priorat.*

residues. Two different laboratories do the testing. The estate is quite isolated and surrounded by forest, otherwise being totally free from residues is not obvious even for an organic estate.

### Domaine Clos Tue-Bœuf, Cheverny
Winemaker Thierry Puzelat calls himself natural and he tries to stay as far away from sulphur, filtration, fining and so on, as possible. Interesting, and sometimes very unusual, whites and reds from Cot.

## And some more recommendations from the Loire Valley
Clos de la Briderie, Touraine
Domaine de l'Ecu, Muscadet
Domaine Pierre Breton, Bourgueil
Château Yvonne, Saumur and Saumur-Champigny
Marc Angeli, Ferme de la Sansonnière, Anjou
Domaine FL, Anjou and Savennières
Domaine Pithon-Paillé

## Alsace

### Domaine Marcel Deiss
Certified Demeter since 1988. Domaine Deiss is a pioneer and a forerunner of organic and biodynamic viticulture in Alsace. Owner and winemaker Jean-Michel Deiss have 27 hectares divided into two hundred different plots. Balance and harmony in the vineyard are key words. It is not always easy, he says, to be biodynamic, but you get results. The importance of *terroir* in Alsace is one of his primary concerns.

### Domaine Sipp, Ribeauvillé
Certified organic since 2008. Etienne Sipp gradually became more and more conscious that the road to more complex, structured and fresher wines was through organic viticulture. 'Wine is culture, not just something you drink with your meal', he says. 'We must respect the wine, the person drinking it and the environment'.

### Domaine Humbrecht, Pfaffenheim, Alsace

Certified Demeter. The family Humbrecht in Pfaffenheim has always been organic and in 2001 Pierre-Paul Humbrecht met biodynamic consultant Pierre Masson and was immediately convinced that this was the right way to make wine. Now Pierre-Paul's son Marc is in charge of winemaking and as passionate about biodynamic agriculture as his father. One result, they both say, is that the vines are much more resistant to diseases.

### Domaine Pierre Frick, Alsace

Certified Demeter since 1981. 'To farm according to the biodynamic principals changes the vineyard', says Jean Pierre Frick. 'It is an immense difference. The vines are more resistant and are less prone to attacks of grey rot'. He discovered biodynamics with his father already in the 1970s. It was an eye-opener; he remembers, a new way to work with nature.

### Domaine Marc Kreydenweiss, Alsace

Certified Biodyvin since 1995. Marc Kreydenweiss is a prominent figure in biodynamics in France. He is also one of the most skilled winemakers in Alsace. With low yields in his well-kept vineyards he manages to obtain a rare complexity and a depth in his wines. He also has a vineyard in the southern Rhône Valley, making excellent but atypical wines.

## And some more recommendations from Alsace:

Domaine Zind-Humbrecht
Domaine Barmes Buecher
Domaine Sylvie Spielmann
Domaine Albert Mann
Domaine Kuentz Bas
Domaine André Ostertag
Domaine Josmeyer
Domaine Gustave Lorentz, Bergheim

# Champagne

### Champagne Fleury, Côte de Bar

Certified Demeter since 1992. The family Fleury started very early with biodynamics. The 13 hectares in Côte de Bar, the southern part of Champagne, is now looked after by sister and brother Morgane and Jean-Sébastien Fleury. 'The most important thing is to be observant all the time', they both say. 'But it also important', they add, 'not to hide behind the label. The quality of the wine needs to be *irreproachable*'.

### Champagne Marguet, Ambonnay

Certified organic since 2008. 'Since we stopped spraying with chemicals our wines taste different, in a positive way', says Benoît Marguet, the owner and winemaker. He has a bit over thirteen hectares and works with low yields, barrel-fermented wine and low dosage.

### Champagne Larmandier-Bernier, Vertus

Certified organic since 2008. The quality of the wine starts in the vineyard, so of course you need to work as naturally as possible, says Pierre Larmandier. 'With mechanical ploughing and low yields our vines do not need any fertilisers. It takes a bit more work, but you taste the *terroir* in our wines'.

### Champagne Jacquesson, Dizy

Champagne Jacquesson has worked as environmental friendly for many years and started their conversion to be certified in 2010. It is a bit risky to be organic in Champagne, says owner and winemaker Jean-Hervé Chiquet, 'definitely more so than close to the Mediterranean. But it is important for the quality. For instance, it is good to have grass growing in the vineyard as it competes with the vines and forces their roots to go deep down in the soil to search for minerals'.

### Champagne Leclerc-Briant, Epernay

Demeter certified. The house owns seven hectares of mainly premier cru vineyards and buys additional grapes from around fifteen hectares. A Champagne Brut Blanc de Noir with a lot of character, slightly oxidised. A different style of champagne.

*Harvesting falanghina grapes at the Ocone winery in Campania.*

## And some more recommendations from Champagne:

Champagne Leclapart
Champagne de Sousa
Champagne Alain Thienot
Champagne Joseph Perrier
Champagne J. de Telmont

## Bordeaux

### Château Pontet-Canet, Pauillac

Certified Biodyvin since 2011. One of the rare 'big and famous' among Bordeaux chateaux to be biodynamic, much thanks to winemaker Jean-Michel Comme. But owner Guy Tesseron and his daughter Mélanie were easily persuaded. They like the idea of being natural. 'With biodynamic you protect instead of cure', says Mélanie, 'and you get a better fruit in the wine. And a fruity wine does not need any makeup', she adds.

### Château Guiraud, Sauternes

Certified organic since 2011. A big Sauternes premier grand cru classé with over one hundred hectares of vines. 'The wines gain in elegance and cleanness', says Xavier Planty, winemaker and one of the owners. And, not least important for a Sauternes château, noble rot tends to arrive more easily.

### Château Moulin du Cadet, Saint Emilion

Certified Biodyvin since 2004. Owner and winemaker Pierre Blois really believes that biodynamic viticulture affects the characteristics of the wine. 'It gives the wine a livelier acidity and a certain saltiness', he says. 'The wines are full-bodied without being heavy'.

### Clos Puy Arnaud, Castillon-Côtes de Bordeaux

Certified Biodyvin since 2007. Thierry Valette, owner and winemaker, recalls his conversion to biodynamics: 'It was the beginning of an extraordinary adventure. After five years it was like an explosion in the vineyard. The soil changed colour, turned red and more alive, flowers started to grow and we had a balance between quality and vigour, it was spectacular. And I had a lower pH in the wine, which means more stable wine and less need for sulphur'.

## And some more recommendations from Bordeaux:

Château Le Puy, Francs Côtes de Bordeaux
Château Fonroque, Saint Emilion
Château La Graves, Fronsac, Paul Barre
Vieux Château Champs de Mars, Castillon-Côtes de Bordeaux
Château de la Dauphine, Fronsac
Château Lapeyronie, Castillon-Côtes de Bordeaux
Château Climens, Barsac
Château Palmer, Margaux

## Burgundy

### Domaine Rossignol-Trapet, Gevrey-Chambertin

Certified Demeter. It is important to be certified, says Nicolas Rossignol: 'it shows that we are a 100% biodynamic and not only when the weather allows it'. It was already in the late 1990s that the family Rossignol realised that the use of synthetic products were not sustainable. They wanted something more environmentally friendly and choose biodynamics. 'It is the only method that takes into account the whole vine, from the roots to the leaves', says Nicolas. 'It makes the vines stronger and more disease resistant'.

### Domaine de la Boissonneuse, Chablis

Certified Demeter since 2005. Boissonneuse is a small, separate vineyard that belongs to Domaine Jean-Marc Brocard in Préhy. It is run by Julien Brocard, son of Jean-Marc. 'Organic', says Julien, 'is to avoid synthetic products, but biodynamics look for the reasons behind diseases and deficits. That appeals to me'.

### Domaine les Temps Perdus, Clotilde Davenne, Chablis

Clotilde Davenne worked for many years as a winemaker for Jean-Marc Brocard. Now she has her own wine estate, not so far away. All of

her wines, ranging from generic Bourgogne to Chablis Grand Cru, are worth trying out.

### Domaine de la Vougeraie, Premeaux Prissey

Owned by famous family company Boisset Vougeraie produces elegant red wines with a very reasonable price tag.

### Domaine de la Soufrandière and Bret Brothers, Mâcon

Two talented brothers, Jean-Guillaume and Jean-Philippe Bret took over the La Soufrandière family estate in southern Burgundy in the early 2000s when it left the local cooperative. Certified Biodynamic Demeter since the early 2000s. Bret Brothers is the brand they use for their *négoce* wines (bought in grapes). Their excellent wines all have a steely, elegant acidity.

### Domaine Leflaive, Puligny-Montrachet

Anne-Claude Leflaive is one of the most highly respected biodynamic producers in France. She converted one of the most prestigious Burgundy *domaines,* Domaine Leflaive, to biodynamic farming after extensive trials and tastings.

### Domaine Dominique et Catherine Derain, Saint Aubin

Dominique makes his own personal style of wine. Not everyone likes this somewhat 'unpolished' and very natural style. But that does not bother Dominique. He has no problem selling his Saint Aubin wines and his very drinkable, unpretentious Allez Goûtons made from the Aligoté grape.

## And some more recommendations from Burgundy:

### Chablis

Alice et Olivier de Moor, Chablis
Domaine Laroche, Chablis
Domaine Christian Moreau Père et Fils, Chablis

*Terraced vineyards in Priorat.*

## *Côte d'Or*

Domaine Leroy
Domaine Prieuré Roch
Domaine Pierre Morey
Domaine Bonneau du Martray
Domaine Comtes Lafon
Domaine Dujac
Domaine Anne Parent
Domaine Thierry Mortet
Domaine Trapet Père & Fils
Domaine Marquis d'Angerville
Domaine Vincent Girardin
Domaine Amiot-Servelle
Domaine du Château de Puligny-Montrachet
Domaine de Chassorney
Domaine Jean Jacques Confuron
Domaine du Comte Liger-Belair
Domaine Remi Jobart
Domaine Lucien Muzard et Fils
Domaine de la Romanée Conti
Domaine René Bouvier
Domaine Henri et Gilles Buisson
Domaine Jean-Noël Gagnard

## *Beaujolais*

Domaine Jean Foillard, Villié-Morgon
Domaine Jean-Paul Thevenet, Villié-Morgon

# Italy

### *Fattoria Poggerino, Radda in Chianti*

Certified organic. Converting to biodynamics. Being organic is now a life style for us, says winemaker Piero Lanza who runs Poggerino with his sister Benedetta. They have eleven hectares. To have quality wines, Piero works hard in the vineyard. The biodynamic preparations help, he thinks. For instance, 500 is a concentrate of energy, he says.

## Gulfi, Sicily

Certified organic. Gulfi is a young vineyard on the south east side of Sicily. It was created in 1998 by the family Catania. 'The land here had never been treated with synthetic products so it felt quite natural to start growing organically,' says Matteo Catania. 'We did not think of it as a sales strategy but as a matter of course'.

## Azienda Agricola Cos, Sicily

COS is located in the southeast of Sicily, ten kilometres from Vittoria, home to Cerasuolo di Vittoria, so far the only DOCG on the island. COS has been run biodynamically for several years. Giusto Occhipinti makes his own teas of nettles and other herbs and plants, but he buys some of the biodynamic preparations. 'Some are hard to prepare yourself and it is not always easy to get hold of the biodynamic ingredients you need'.

## Colombaia, Chianti

A small (four hectares) Chianti producer owned and run by husband and wife Dante Lomazzi and Helena Variara. They only work with natural yeasts and with minimal intervention in the cellar which can make in particular their full-bodied white wines a surprise for those who don't know their wines.

## Villa Bellini, Valpolicella

Villa Bellini is a small estate run by talented Cecilia Trucchi. She designs her own artistic wine labels.

## La Cappuccina, Soave

This beautiful estate with its own private church from the thirteenth century is run by the Tessari family. They produce elegant and aromatic Soave and they also experiment with old and almost forgotten grape varieties.

## Giovanni Menti, Gambarella

Stefano Menti is the owner and wine maker of this interesting estate of 7.5 hectares in Gambarella, a region east of Soave. He makes a wide range of wine for such a small estate. Some of it is very 'natural' in style and spirit. Excellent quality with lots of originality, including sweet white wine from dried grapes.

### Renzo Marinai, Tuscany

A charming vineyard of six hectares close to Panzano in Chianti Classico. Mozart is played in the wine cellar.

## And some more recommendations from Italy:

### Piedmont

Punset, Barbaresco
Erbaluna, Barolo
Cascina Corte

### Veneto

Fasoli Gino, Soave
Azienda Agricola Musella, Valpolicella

### Tuscany

Il Paradiso di Frassina, Montalcino
Fonterenza, Montalcino
Badia a Coltibuono, Chianti
Tenuta di Poggio, Chianti
San Polino, Montalcino
Poggiopaoli, Morellino di Scansano
Cortona Vino Stefano Amerighi
La Castellaccia

### Marche

Umani Ronchi

### Campania

Ocone

*A vineyard in Stellenbosch.*

*Puglia*
Le Fabriche

## Spain

### *Albet i Noya, Catalonia*

Certified organic. Albet i Noya in Catalonia launched early into organic farming. They made their first organic wine already at the end of the 1970s. The family stresses the importance of harvesting healthy, non-damaged grapes. Also the hygiene in the cellar is important. It helps to avoid using too much sulphur.

### *And some more recommendations from Spain:*

Alvaro Palacios, Priorat
Mas Comtal, Penedès, Avinyonet del Penedès
Mas Igneus, Priorat
Parés Baltà, Pacs del Penedès
Ijalba, Rioja
Bodega Casa de la Ermita, Jumilla
Telmo Rodriguez, Rioja

## Austria

### *Nicolaihof, Wachau*

Certified Demeter. Nikolaus and Christine Saachs are pioneers in biodynamic agriculture. They were inspired to start with it by reading the books of Rudolf Steiner.

### *Meinklang, Neusiedlersee*

Certified Demeter. Werner and Anneliese Meinklang had been organic for ten years when they took on a new challenge, to convert to biodynamics. They are unusual in that they really are a true biodynamic farm. They do not have only vines but also 300 Angus cows and fields with flowers, other plants and grain.

## Weingut Sepp Moser

Certified Demeter since 2009. Winemaker Nikolaus Moser says it was difficult in the beginning, especially the first year. But after that he noticed the balance in the wine. 'I get a better tannin structure and acidity with biodynamics', he says. 'Especially my red wines from Zweigelt are so much better'.

# Germany

## Dr Bürklin-Wolf, Pfalz

Certified Biodyvin since 2005. Ploughing is done by horse and the biodynamic compost as well as several of the preparations are made on site.

# California

## Fetzer Vineyards and Bonterra Vineyards

Certified Demeter. The organic wine movement in California started in the 1960s. The pioneers were mostly in Mendocino County. One of them was Paul Dolan who for many years worked at Fetzer Winery where he introduced both organic and biodynamic farming and a sustainable philosophy in general. Bonterra is a brand founded by Fetzer in 1990 and which today is California's largest exporter of organically grown wines. Both brands are now owned by the big Chilean group Concha y Toro.

## Paul Dolan Vineyards, Mendocino County

Paul Dolan has left Fetzer but he continues to make organic and biodynamic wines on his properties. He is pleased that organic viticulture is growing in California. 'I think it is because producers see the opportunity they have to make better wines through organic

farming', he says. 'I think that organic wines are not only of higher quality, but they also provide a transparency for consumers'. He appreciates the fact that organic wine producers in California have chosen to use only 100 % organic grapes, although in practice they could add up to 30% non-organic grapes if they wanted to (according to the rules) and still be able to call their wines 'made with organic grapes'. The food industry, he says, typically takes advantage of such an opportunity.

## La Rocca, California

Certified organic through NOP. One of the world's organic pioneers. Philip La Rocca has harvested organic fruit and grapes for forty years. It started with the first organic apples in California in 1974. His wines are certified organic so they are made without any sulphur at all which is the way an organic wine should be, says Philip.

## Frey Vineyards, Mendocino County

Frey is certified organic through NOP since 1980. According to owner Alexander Frey the family has a long experience of making wines without sulphur. It can be tricky, he says, but he does not want to change the rules.

## And some more recommendations from California:

Frog's Leap Wine Cellars
Robert Sinskey, Napa (Demeter)
Tablas Creek Vineyard, Paso Robles
Bonny Doon (Demeter)
DeLoach Vineyards (Demeter)
Grgich Hills Estate (Demeter)
Joseph Phelps Vineyards (Demeter)
Hedges Estate (Washington) (Demeter)
Benziger (Demeter)

*Steep slopes in the Wachau region along the River Danube, Austria.*

## Chile

### Cono Sur, Chile

A bicycle symbolises the environmentally friendly work at Viña Cono Sur. The winery is part of the big Concha y Toro group but they operate independently and with their own wine makers. Climatic conditions here in Chile make it easy to be organic, says wine maker Matías Ríos. To get rid of certain insects they have geese strutting around in the vineyard. In 2007 Cono Sur received for the first time a certificate called Carbon Neutral Delivery Status. This certificate means that the winery compensates for its carbon emissions during bottle transports. Cono Sur has 1500 hectares and part of the estate is certified organic.

### Emiliana, Casablanca, Chile

Certified Demeter. Emiliana is a big estate with over 1000 hectares of vines and a pioneer in organics. It started already in 1998 with the help of biodynamic consultant Alvaro Espinoza. One big advantage with organics is not having to look like a cosmonaut when you do your treatments, says Nicolás Pollman Loosli at Emiliana.

## Argentina

### Domaine Jean Bousquet, Uco Valley, Mendoza

Jean Bousquet came to Mendoza from the Languedoc in France in 1997. Today the company is the largest exporter of organic Argentinian wine. 'We export 90% of our production', says Guillaume Bousquet. 'We don't sell our wines because they are organic though', he adds, 'the most important thing is still to make good wines. The climate is dry, which helps. The biggest climatic problem is hail'.

### Zuccardi, Mendoza

Zuccardi is a big family company in Mendoza, growing a wide variety of grapes. Part of the estate is organic.

*A vineyard in Mendoza with nets to protect from hail damage, at the foot of the Andes.*

### Colomé, Salta

Certified Demeter. The vines are grown at very high altitudes, up to over 3000 metres.

### Bodega Krontiras, Mendoza

A recently founded Greek-owned winery still in the process of building the cellars. Organic certification 2008 and under conversion to Demeter biodynamics.

## South Africa

### Reyneke Wines, Stellenbosch

Certified Demeter. Johan Reyneke Jr. took over the family estate of forty hectares in 1998 and began to develop the production of wines. He also went from conventional farming to organic farming and subsequently became one of the first in South Africa to produce biodynamic wines.

### Avondale, Stellenbosch

A relatively large estate of 100 hectares. Certified organic by Control Union, one of Europe's largest certification organisations. Avondale was one of the first estates in South Africa that became BWI certified. BWI stands for the Biodiversity in Wine Initiative and was founded by the Botanical Society of South Africa. BWI aims to encourage farmers to plant and maintain indigenous trees, plants and flowers in and around the farms.

### Stellar Organics, Olifants River

Stellar Organics is located in the hot and arid Olifants River that has less problems with fungal diseases than Stellenbosch. The winery is certified organic by Control Union and the USDA. Stellar Organics has an interesting range of wines with no added sulphur.

## Australia

Cullen Wines, Margaret River, Western Australia
Frogmore Creek Vineyards, Tasmania
Temple Bruer Wines, South Australia
Paxton Vineyards, MacLaren Vale
Robinvale Wines, Victoria


## New Zealand

Seresin Estate, Marlborough (biodynamic)
Felton Road, Central Otago (biodynamic)
Kawarau Estate Vineyard, Central Otago
Millton Vineyard, Gisborne (biodynamic)

# Index

acidity reduction 169
additives 151f, 155–82
Agence Bio 103
albumin 175
Alko 201
allergenic ingredients 157
ammonium bisulphite 159
ammonium sulphate 162
ascorbic acid 159

bentonite 175
beta glucanase 177
Biodiversity in Wine Initiative (BWI) 198
biodynamic wines 111-31
Biodyvin 126, 141
Bio Federal 106
Bordeaux mixture 80
botrytis 81
Bouchet, François 124
Bourguignon, Claude 65, 74
Brettanomyces 151

calcium carbonate 169
carbon dioxide 74, 209, 220
carbon
—, active 172
— emissions 212
carbon-neutral 213
casein 172

cellulose plates 176
centrifuging 177
certification 101, 109
Certified Sustainable Wine of Chile 197
chaptalisation 164
chitosan 170
cold stabilisation 177
compost 71
contact treatment 64
conversion 61, 103
copper 64, 80, 91, 94
copper citrate 179
copper sulphate 179
cork 217
Cousinié Method 190
cover crop 72
cryo-extraction 167

Demeter 124, 126, 140
diammonium phosphate 162
dimethyl dicarbonate (DMDC) 171
diseases, fungal 79, 98
distillation, vacuum 166
dynamisation 116

Ecocert 102
electrodialysis 169, 178
enzymes 161
EU regulation
— organic farming 30, 40, 61, 101

*Chianti Classico, Tuscany.*

— organic wine 21, 62
Fairtrade 200
fermentation, malolactic 167
fertiliser 68
—, artificial 39, 113
filter tangential 176
filtration 176
fining 174
fish glue 176
*flavescence dorée* 87
Forest Stewardship Council 218

gelatine 175
Generational Farming Program 200
genetically modified (GM) products 99
grape worm 81
grape must, concentrated 166
Gum Arabic 174

Haber-Bosch method 35
hedges 74
horsetail 89, 95, 116, 142,
hybrids 98

insects 74
insecticide 35, 83
Integrated Production of Wine (IPW)
   197
ion exchange 178
irrigation 67, 219

Joly, Nicolas 40, 43, 143, 237

kieselguhr 176

labour costs 58
L'Association des Vins Naturels 203
lysozymes 171

malic acid 168
mannoproteins 178
Masson, Pierre 124
metatartaric acid 178
mildew
—, downy 36, 79
—, powdery 36, 80

monoculture 39, 64, 77
mycelium 72, 119

National Organic Program 105
natural wines 203–7
nettles 89, 95, 120
nitrogen 36, 68, 70, 74
nitrogen, phosphorus and potassium
   (NPK) 67f

oak barrels 180
oak chips 180
Öpul Integrierte Produktion 194
osmosis, reverse 167

pasteurisation, flash 171
PET bottles 215
phosphorus 68, 70
photosynthesis 67, 70
phylloxera 36, 98
planting calendar 122
— *The Maria Thun Biodynamic
   Calendar* 122
plant stimulator 81, 95
ploughing 74
—, horse-drawn 251
Podolinsky, Alex 124
polyvinyl polypyrrolidone (PVPP) 172
potassium 68, 71
potassium bicarbonate 169
potassium bisulphite 159, 170
potassium metabisulphite 161
potassium tartrate 169
preparations
—, biodynamic 112f, 121
—, natural 89
*préparations naturelles peu préoccupantes*
   (PNPP) 89, 121
pyrethrin 35, 87

reduction 179
roots 67
rotenone 35, 87

screw cap 217
sexual confusion 81, 83f

silica (silicon dioxide) 176
soil erosion 69, 72
solar panels 218
sorbic acid 172
spider, red 84
spraying 21, 61f, 79, 87, 91–95, 97, 101,
    185f, 190
stabilisation 170
Steiner, Rudolf 39, 111
sulphur 64, 80, 146, 158, 181, 204, 206
sulphur dioxide 159, 170
Sustainable Wine-growing New Zealand
    (SWNZ) 198
Sustainable Winegrowing Program 194
Sustaining Success 200
Systembolaget 201
systemic treatment 64, 94

tannins 179
tartaric acid 168
temperature control 161

Terra Vitis 193
thiamine 161
Thun, Maria 122, 124
— *The Maria Thun Biodynamic
    Calendar* 122
trace elements 68
trichoderma 81, 95
Tyflo 193

US Department of Agriculture (USDA)
    105

ViniVeri 205
VinNatur 205
vins nature: *see* natural wines

weather 95

yeast 126, 152, 163, 204
— cell walls 162
yield 58

# When Wine Tastes Best App

A fun and easy way to discover
when wine will taste its best.

- Look ahead to key dates – search
  by month
- Swap between day and week view –
  see when wine tastes best at a glance
- Automatically adjusts to your
  time zone
- Try it for free – get a whole year's
  data with in app purchase

# Biodynamic Gardening Calendar App

A quick, easy way to look up the key
sowing and planting information
found in the original *Maria Thun
Biodynamic Calendar*.

- Filter activities by the time types of
  the crops you're growing
- Automatically adjusts to your
  time zone
- Plan ahead by day, week or month
- Available in English, German and Dutch

# Recommended reading

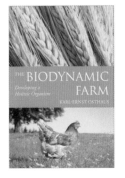

THE BIODYNAMIC FARM
Developing a Holistic Organism
KARL-ERNST OSTHAUS

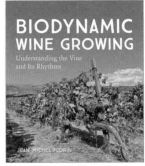

BIODYNAMIC WINE GROWING
Understanding the Vine and Its Rhythms
JEAN-MICHEL FLORIN

The Biodynamic Orchard Book
Ehrenfried Pfeiffer and Michael Maltas

Companion Plants and How to Use Them
Helen Philbrick & Richard B. Gregg

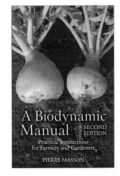

A Biodynamic Manual SECOND EDITION
Practical Instructions for Farmers and Gardeners
PIERRE MASSON

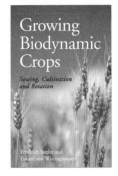

Growing Biodynamic Crops
Sowing, Cultivation and Rotation
Friedrich Sattler and Eckard von Wistinghausen

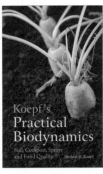

Koepf's Practical Biodynamics
Soil, Compost, Sprays and Food Quality. Herbert H. Koepf

THE Maria Thun BIODYNAMIC CALENDAR 2021
OVER 100,000 COPIES SOLD

Pfeiffer's Introduction to Biodynamics
Ehrenfried Pfeiffer

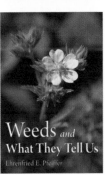

Weeds and What They Tell Us
Ehrenfried E. Pfeiffer

What's so Special about Biodynamic Wine?
35 Questions & Answers for Wine Lovers
Antoine Lepetit de La Bigne

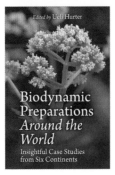

Edited by Ueli Hurter
Biodynamic Preparations Around the World
Insightful Case Studies from Six Continents

florisbooks.co.uk

Floris
Books

For news on all the latest books, and to get
exclusive discounts, join our mailing list at:

florisbooks.co.uk/mail/

*And* get a FREE book
with every online order!

*We will never pass your details to anyone else.*